THE PLANETS

Also by Dava Sobel

To Father
Galileo's Daughter
Longitude

A studious observer of the Moon, planets and comets, 'England's Leonardo' Robert Hooke (1635–1703) drew this design for a tubeless telescope to improve his view.

THE
PLANETS

DAVA SOBEL

FOURTH ESTATE · *London*

First published in Great Britain in 2005 by
Fourth Estate
An imprint of HarperCollins*Publishers*
77–85 Fulham Palace Road
London W6 8JB
www.4thestate.co.uk

Originally published in the United States in 2005 by Viking

'Pluto' and 'Venus' quoted from *The Planets: A Cosmic Pastoral*
by Diane Ackerman, published by William Morrow.
Reprinted by permission of the poet.
Carl Sagan quotation from *The Cosmic Connection:
An Extra-terrestrial Perspective*, produced by Jerome Agel.
Reprinted by permission of Jerome Agel.
'The Literate Farmer and the Planet Venus' quoted from
The Poetry of Robert Frost, edited by Edward Connery Latham
and published by Jonathan Cape. Reprinted by permission of
The Random House Group Ltd.

6

A catalogue record for this book
is available from the British Library

ISBN 1-85702-850-3

Set in PostScript Linotype Giovanni Book
with Castellar and Spectrum display by
Rowland Phototypesetting Ltd, Bury St Edmunds, Suffolk
Printed and bound in Great Britain by
Clays Ltd, St Ives plc

Dedicated with worldfuls of love to my big brothers,
Michael V. Sobel, MD,
who named our family cat Captain Marvel,
and Stephen Sobel, DDS,
who bunked with me at Space Camp.

At night I lie awake
in the ruthless Unspoken,
knowing that planets
come to life, bloom,
and die away,
like day-lilies opening
one after another
in every nook and cranny
of the Universe. . . .

– Diane Ackerman,
from *The Planets: A Cosmic Pastoral*

In all the history of mankind, there will be only one generation that will be first to explore the Solar System, one generation for which, in childhood, the planets are distant and indistinct discs moving through the night sky, and for which, in old age, the planets are places, diverse new worlds in the course of exploration.

– Carl Sagan, from *The Cosmic Connection: An Extra-terrestrial Perspective*

CONTENTS

THE PLANETS

1

MODEL WORLDS

My planet fetish began, as best I can recall, in third grade, at age eight – right around the time I learned that Earth had siblings in space, just as I had older brothers in high school and college. The presence of the neighbouring worlds was a revelation at once specific and mysterious in 1955, for although each planet bore a name and held a place in the Sun's family, very little was known about any of them. Pluto and Mercury, like Paris and Moscow, only better, beckoned a childish imagination to ultra-exotic utopias.

The few sure facts about the planets suggested fantastic aberrations, ranging from unbearable extremes of temperature to the warping of time. Since Mercury, for example, could circle the Sun in only eighty-eight days, compared to Earth's 365, then a year on Mercury would whiz by in barely three months, much the way 'dog years' packed seven years of animal experience into the dog owner's one, and thereby accounted for the regrettably short lives of pets.

Every planet opened its own realm of possibility, its own version of reality. Venus purportedly hid lush swamps under its perpetual cloud cover, where oceans of oil, or possibly soda water, bathed rain forests filled with yellow and orange plant life. And these opinions issued from serious scientists, not comic books or sensational fiction.

The limitless strangeness of the planets contrasted sharply with their small census. In fact, their nine-ness helped define them as a group. Ordinary entities came in pairs or dozens, or quantities ending in a five or a zero, but planets numbered nine and nine only. Nine, odd as outer space itself, could nevertheless be counted on the fingers. Compared to the chore of memorizing forty-eight state capitals or significant dates in the history of New York City, the planets promised mastery in an evening. Any child who committed the planets' names to memory with the help of an appealing nonsense-sentence mnemonic – 'My very educated mother just served us nine pies' – simultaneously gained their proper progression outward from the Sun: Mercury, Venus, Earth, Mars, Jupiter, Saturn, Uranus, Neptune, Pluto.

The manageable sum of planets made them seem collectable, and motivated me to arrange them in a shoe-box diorama for the science fair. I gathered marbles, jacks balls, ping-pong balls, and the pink rubber Spaldeens we girls bounced for hours on the sidewalk, then I painted them with tempera, and hung them on pipe cleaners and string. My model (more like a doll's house than a scientific demonstration) failed to give any real sense of the planets' relative sizes or the enormous distances between them. By rights I should have used a basketball for Jupiter, to show how it dwarfed all the others, and I should have mounted everything in a giant carton from a washing machine or a refrigerator, the better to approximate the Solar System's grandiose dimensions.

Fortunately my crude diorama, produced with a complete lack of artistic skill, did not kill my beautiful visions of Saturn suspended in the perfect symmetry of its spinning rings, or the mutating patterns on the Martian landscape, which were attributed, in scientific reports of the 1950s, to seasonal cycles of vegetation.

After the science fair, my class staged a planets play. I got the part of 'Lonely Star' because the script called for that character to wear a red cape, and I had one, left over from a Hallowe'en costume. As Lonely Star, I soliloquized the Sun's wish for companionship, which the planet-actors granted by joining up with me, each in a speech admitting his own peculiarities. The play's most memorable performances were delivered by 'Saturn', who twirled two hula-hoops while reciting her lines, and 'The Earth', plump and self-conscious, yet forced to announce matter-of-factly, 'I am twenty-four thousand miles around my middle.' Thus was the statistic of the earth's circumference indelibly impressed upon me. (Note that we always said 'the earth', in those days. 'The earth' did not become 'Earth' until after I came of age and the Moon changed from a nightlight to a destination.)

My role as Lonely Star helped me appreciate the Sun's relationship to the planets as parent and guide. Not for nothing is our part of the universe called the 'Solar System', in which each planet's individual makeup and traits are shaped in large measure by proximity to the Sun.

I had omitted the Sun from my diorama because I hadn't understood its power, and besides, it would have posed an impossible problem of scale.* Another reason for leaving out the Sun, and likewise the Moon, was the bright familiarity of both objects, which seemed to render them regular components of Earth's atmosphere, whereas the planets were glimpsed only occasionally (either before bedtime or in a still-dark, early-morning sky), and therefore more highly prized.

---

* In his ingenious pamphlet, 'The Thousand-Yard Model, or, The Earth as a Peppercorn', Guy Ottewell guides the construction of a scale model Solar System using a bowling ball for the Sun. The eight-thousand-mile-wide Earth, here reduced to a peppercorn, takes its rightful place seventy-eight feet (!) from the bowling ball.

On our class trip to the Hayden Planetarium, we city kids saw an idealized night sky, liberated from the glare of traffic signals and neon lights. We watched the planets chase each other around the heavens of the dome. We tested the relative strength of gravity with trick scales that told how much we'd weigh on Jupiter (four hundred pounds and more for a normal-sized teacher) or Mars (featherweights all). And we gawked at the sight of the fifteen-ton meteorite that had fallen from out of the blue over Oregon's Willamette Valley, posing a threat to human safety that few of us had thought to fear.

The Willamette meteorite (still on permanent display at what is now the Rose Center for Earth and Space) was said to be, incredibly, the iron-nickel core of an ancient planet once in orbit around the Sun. That world had shattered somehow, several billion years back, setting its fragments adrift in space. Chance had nudged this particular piece toward the Earth, where it hurtled down to the Oregon ground at tremendous speed, burning up from the heat of friction, and hitting the valley floor with the impact of an atom bomb. Later, as the meteorite lay still over aeons, the acidic rains of the Pacific northwest chewed large holes in its charred and rusted hulk.

Here was a primal scene to upset my innocent planet ideas. This dark, evil invader had no doubt consorted in space with hordes of other stray rocks and metal chunks that might strike Earth at any moment. My Solar System home, till that moment a paragon of clockwork regularity, had turned into a disorderly, dangerous place.

The launch of *Sputnik* in 1957, when I was ten, scared me to death. As a demonstration of foreign military strength, it gave new meaning to the school-wide air raid drills in which we crouched under our desks for pretended safety, our backs to the windows. Clearly we still had more to dread from angry fellow humans than from wayward space rocks.

All through my teens and twenties, as the country realized the young president's dream of a rocket to the Moon, clandestine rockets in missile silos kept collective nightmares alive. But by the time the Apollo astronauts brought back their last batch of Moon rocks in December 1972, peaceful, hopeful spaceships had landed also on Venus and Mars, and another, the US *Pioneer 10*, was en route to a Jupiter flyby. Throughout the 1970s and 1980s, hardly a year passed without an unmanned excursion to another planet. Images radioed home to Earth by robot explorers painted detail upon detail on the planets' long-blank faces. Whole new entities came to light, too, as spacecraft encountered literally dozens of new moons at Jupiter, Saturn, Uranus and Neptune, as well as multiple rings around all four of those planets.

Even though Pluto remained unexplored, deemed too

distant and too difficult to visit, its own unexpected moon was discovered accidentally in 1978, through careful analysis of photographs taken by ground-based telescopes. Had my daughter, born in 1981, attempted her own diorama of the revised and expanded Solar System when she turned eight, she would have needed handfuls of jellybeans and jawbreakers to model the many recent additions. My son, three years her junior, might have opted to model his on our home computer.

Despite the increased population of the Solar System, its planets stayed stable at nine, at least until 1992. That year, a small, dark body, independent of Pluto, was detected on the Solar System's periphery. Similar discoveries soon followed, until the total number of diminutive outliers grew to seven hundred over the ensuing decade. The abundance of mini-worlds made some astronomers wonder whether Pluto should continue to be regarded as a planet, or reclassified as the largest of the 'trans-Neptunian objects'. (The Rose Center has already excluded Pluto from the planetary roll call.)

In 1995, only three years after the first of Pluto's numerous neighbours was found, something even more remarkable came to light. It was a bona fide new planet – of another star. Astronomers had long suspected that stars other than the Sun might have their own planetary systems, and now the first one had surfaced at 51 Pegasi, in the constellation of the flying horse. Within months, other 'exoplanets' – as the newly discovered extra-solar planets were quickly dubbed – turned up at stars such as Upsilon Andromedae, 70 Virginis b, and PSR 1257+12. At least 180 additional exoplanets have since been identified, and refinements in discovery techniques promise to uncover many more in the near future. Indeed the number of planets in our Milky Way Galaxy alone may far exceed its complement of one hundred billion stars.

My old familiar Solar System, once considered unique, now stands as merely the first known example of a popular genre.

As yet, no exoplanets have been imaged directly through a telescope, so their discoverers are left to imagine what they look like. Only their sizes and orbital dynamics are known. Most of them rival giant Jupiter in heft, because large planets are easier to find than small ones. Indeed, the existence of exoplanets is deduced from their effect on their parent star: Either the star wobbles as it yields to the gravitational attraction of unseen companions, or it periodically dims as its planets pass in front and impede its light. Small exoplanets, the size of Mars or Mercury, must also orbit distant suns, but, being too tiny to perturb a star, they elude detection from afar.

Already planetary scientists have appropriated the name 'Jupiter' as a generic term, so that 'a jupiter' means 'a large exoplanet', and the mass of an extremely large exoplanet may be quantified as 'three jupiters' or four.

In the same fashion, 'an earth' has come to represent the most difficult, most desirable goal of today's planet hunters, who are devising ways to probe the Galaxy for petite, fragile spheres in the favoured shades of blue and green that hint at water and life.

Whatever daily concerns dominate our minds at the dawn of the present century, the ongoing discovery of extra-solar planetary systems defines our moment in history. And our own Solar System, rather than shrink in importance as one among many, proves the template for comprehending a plethora of other worlds.

Even as the planets reveal themselves to scientific investigation, and repeat themselves across the universe, they retain the emotional weight of their long influence on our lives, and all that they have ever signified in Earth's skies. Gods of old, and demons, too, they were once – they still are – the sources of an inspiring light, the wanderers of night, the far horizon of the landscape of home.

2

GENESIS

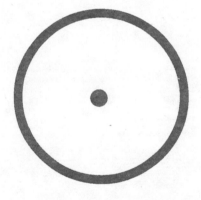

'In the beginning, God created the heaven and the earth,' the first book of the Bible recounts. 'And the earth was a formless void and darkness covered the face of the deep, while a wind from God swept over the face of the waters. Then God said, "Let there be light"; and there was light.'

The energy of God's intent flooded the new heaven and earth with light on the very first day of Genesis. Light's potent good thus pervaded the evenings and the mornings when the seas separated from the dry lands, and the earth brought forth grass and fruit trees – even before God set the sun, moon and stars in the firmament on the fourth day.

The scientific Creation scenario likewise unleashes the universe in a burst of energy from a void of timeless darkness. About thirteen billion years ago, scientists say, the hot light of the 'Big Bang' erupted, and separated itself instantly into matter and energy. The next three minutes of cooling precipitated all the atomic particles in the universe, in the unequal proportions of 75 per cent hydrogen to 25 per cent helium, plus minuscule traces of a few other elements. As the universe expanded exponentially in all directions and continued to cool, it shed no new light for at least a billion years – until it begat the stars, and the stars began to shine.

New stars lit up by pressuring the hydrogen atoms deep within themselves to fuse with one another, yielding helium and releasing energy. Energy fled the stars as light and heat, but helium accumulated inside them, until eventually it, too, became a fuel for nuclear fusion, and the stars melded atoms of helium into atoms of carbon. At later stages of their lives, stars also forged nitrogen, oxygen and even iron. Then, literally exhausted, they expired and exploded, spewing their bounty of new elements into space. The largest and brightest stars bequeathed to the universe the heaviest of elements, including gold and uranium. Thus the stars carried on the work of Creation, hammering out a wide range of raw materials for future use.

As the stars enriched the heavens that had borne them, the heavens gave rise to new generations of stars, and these descendants possessed enough material wealth to build attendant worlds, with salt seas and slime pits, with mountains and deserts and rivers of gold.

In its own beginning, some five billion years ago, the

star that is our Sun arose from a vast cloud of cold hydrogen and old stardust in a sparsely populated region of the Milky Way. Some disturbance, such as the shock wave from a nearby stellar explosion, must have reverberated through that cloud and precipitated its collapse. Widely dispersed atoms gravitated into small clumps, which in turn lumped together, and kept on aggregating in an ever-quickening rush. The cloud's sudden contraction raised its temperature and set it spinning. What had once been a diffuse, cool expanse of indeterminate shape was now a dense, hot, spherical 'proto-solar nebula' on the verge of starbirth.

The nebula flattened into a disk with a central bulge, and there in the heart of the disk the Sun came to light. At the moment the Sun commenced the self-consumptive fusion of hydrogen in the multi-million-degree inferno of its core, the outward push of energy halted the inward gravitational collapse. Over the ensuing few million years, the rest of the Solar System formed from the leftover gas and dust surrounding the infant Sun.

The Book of Genesis tells how the dust of the ground, moulded and exalted by the breath of life, became the first man. The ubiquitous dust of the early Solar System – flecks of carbon, specks of silicon, molecules of ammonia, crystals of ice – united bit by bit into 'planetesimals', which were the seeds, or first stages, of planets.

Even as they assembled themselves, the planets asserted their individuality, for each one amassed the

turn red, and swell in size until it swallows up the planets Mercury and Venus, and melts the surface of the Earth. One hundred million years later, when the Sun has reduced more helium to carbon ash, it will shrug off its outer layers and dispatch them past Pluto. A larger star could resort to carbon burning at that point, but our Sun, a relatively small star by the standards of the universe, will be unable to do so. Instead, it will smoulder as an ember, and shed a fading light on the charred cinder where God once walked among men. This dim future, however, lies so far ahead as to allow the descendants of Adam and Noah ample time to find another home.

The glorious Sun of our time, the planets' progenitor and chief source of energy, embodies 99.9 per cent of the mass in the Solar System. Everything else – all the planets with their moons and rings, plus all the asteroids and comets – account for only .1 per cent. This gross inequity between the Sun and the sum of its companions defines their balance of power, for the universal law of gravity decrees that the massive shall have dominion over the less massive. The Sun's gravity keeps the planets in orbit and also dictates their speeds: the closer they are to the Sun, the faster they go. The Sun, in turn, bends to the will of the concentrated mass of stars at the centre of our Milky Way Galaxy, around which it orbits once every 230 million years, carrying the planets along with it.

Just as they feel the Sun's attraction more or less keenly, according to their distance, so too do the planets partake of the Sun's light and heat. Solar energy diminishes in intensity as it radiates through interplanetary space. And so, while parts of Mercury bake at five hundred degrees, Uranus, Neptune and Pluto remain perpetually deep-frozen. Only in the Solar System's milder middle section, called the habitable zone, have conditions supported the flourishing of 'great whales, and every living creature that moveth, which the waters brought forth abundantly, after their kind, and every winged fowl after his kind: and . . . cattle, and creeping thing, and beast of the earth . . .'

The planets return the favour of the Sun's light by reflecting its rays, and in this manner they pretend to shine, though they emit no light of their own. The Sun is the Solar System's sole light-giving body; all the others glow by reflected glory. Even the full Moon that illumines so many lovely Earthly evenings owes its silvery light to Sunbeams bouncing off the dark lunar soil. The Earth shines just as beautifully when viewed from the Moon, and for the same reason.

The play of mirrored light from Venus, close to the Sun and also closest to Earth, makes that planet appear by far the brightest one to our eyes. Jupiter, though much larger, lies many millions of miles beyond, and therefore pales in comparison in our night sky. The even further worlds of Uranus and Neptune, immense as they are, catch and toss back so little light that Uranus can only occasionally be discerned (as a mere point of light) by the naked eye, Neptune never.

Although Pluto, too, is impossible for us to see without a telescope, other objects on the outskirts of the Solar System can and sometimes do flare into sudden visibility. When disturbed by a chance encounter, an ice-rock denizen from Pluto's depths may be pushed Sunward and transformed from a dull lump into a spectacular *comet*. Basking in the Sun's warmth, the frozen body heats up and throws out a trailing tail of castoff gas and icy dust that sparkles with the Sun's reflection. The brilliance fades and disappears, however, after the comet rounds the Sun and returns to the outer Solar System.*

The visits of comets, long interpreted as signs and wonders, have recently sketched the true extent of the Sun's domain. By tracing the visible parts of comet paths, and extrapolating the rest, astronomers have shown that numerous comets hail from out beyond Pluto's neighbourhood, from a second comet reservoir, hundreds of times further away. Despite their unimaginable distance, these bodies still belong to the Sun, still heed the Sun's gravity, still receive some glimmer of the Sun's light.

---

* Discarded comet dust litters interplanetary space, and when the Earth trundles into a patch of it, the particles that fall through the atmosphere are incinerated, appearing as isolated 'shooting stars' or whole showers of meteors.

Sunlight, which darts through space at the dazzling speed of 186,000 miles per second, takes ages to emerge from the dense interior of the Sun. Light advances only a few miles per year near the Sun's core, where the crush of matter repeatedly absorbs it and impedes its escape. Radiating this way, light may journey for a million years before reaching the Sun's convective zone, there to catch a quick ride up and out on roiling eddies of rising gas. As soon as these eddies release their cargoes of light, they sink back down, to soar again later with more.

The light-emitting, visible surface of the Sun – the photosphere – seethes as though boiling from the constant tumult of energy release. Gas bubbles bursting with light give the photosphere a grainy complexion, marred here and there by pairs of dark, irregularly shaped sunspots, with black centres and grey, graded shading around them like the penumbras of shadows. Sunspots designate areas of intense magnetic activity on the Sun, and their darkness bespeaks their relative coolness of about 4000°K, compared to neighbouring areas at nearly 6000.* The level of solar activity rises and falls in cycles averaging eleven years, and sunspots

---

* Degrees K, for Kelvin, are the same size as degrees Celsius (or centigrade) – almost double the value of Fahrenheit degrees. However the Kelvin scale starts lower, at −273°C, or 'absolute zero', the point at which all motion ceases, and has no upper limit, which makes it useful for describing the temperatures of stars.

mingle, morph, and multiply according to this same schedule. Their number and distribution vary like famine and plenty, from no spots at 'solar minimum', or just a few spots dotting the Sun's high latitudes, to 'solar maximum' five to six years later, when hundreds of them crowd closer to the equator. Although sunspots seem to gather and scud like clouds across the photosphere, really it is the Sun's rotation that carries them around.

The Sun rotates on its axis approximately once a month, in a continuation of the spinning motion it was born in. Being an enormous ball of gas, the Sun spins complexly, in layers of various speeds. The core and its immediate surroundings turn at one rate, as a solid body. The overlying zone spins faster, and above that, the visible photosphere whirls around at several different rates, more quickly at the Sun's equator than near its poles. These combined, contrary motions whip the Sun into a fury, with consequences felt clear across the Solar System.

The 'solar wind', a hot exhalation of charged particles (reminiscent of the 'wind from God'), blows out from the turbulent Sun and keeps up a constant barrage on the planets. Were it not for the protective envelope of Earth's magnetic field, which deflects most of the solar wind, humankind could not withstand the onslaught. From time to time, especially during solar maximum, the steady solar wind is augmented by sudden blasts of higher-energy particles from solar flares on the Sun's surface, or by gargantuan blobs of ejected solar gas. Such outbursts can disable our communications satellites and disrupt power grids, causing blackouts. In milder doses, particles of solar wind trickle into the upper atmosphere near the North and South Poles, initiating cascades of electrical charge that draw curtains of coloured lights across the sky – the so-called Northern and Southern Lights. Other planets also sprout colourful auroras in response to the solar wind, which billows on past Pluto all the way to the heliopause – the undiscovered boundary where the Sun's influence ends.

From Earth, we see the Sun as a blazing circle in the sky, brighter but no bigger than the circumference of the full Moon. The 'two great lights', as the Sun and Moon are described in Genesis, make a matched pair. For although the Moon measures only one four-hundredth the Sun's million-mile diameter, it nevertheless lies four hundred times closer to Earth. This uncanny coincidence of size and distance enables the puny Moon to block out the Sun whenever the two bodies converge on their shared path across Earth's sky.

Approximately once every two years, some narrow swath of Earth – as often as not a godforsaken, all but inaccessible place – is blessed with a total solar eclipse. There, dusk falls and dawn breaks twice on the same day, and the stars come out with the Sun still overhead. Temperatures may drop ten or fifteen degrees at a stroke,

Total Eclipse of

Sir John Herschel (1792–1871) summarised several studies of the Sun in this composite drawing. At upper right he shows a sunspot near the limb (edge) of the photosphere, while in the three other small frames he pays closer attention to individual sunspot

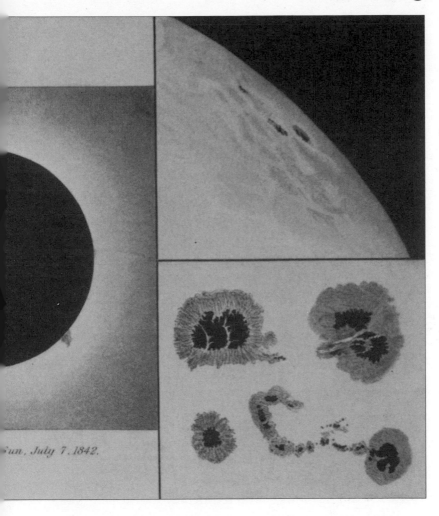

structure. At centre he depicts the Sun's normally invisible corona along with the three prominences he observed during the total solar eclipse of 7 July 1842.

allowing even the most jaded observer to sense the bizarre disorientation that birds and animals share as they hasten to their nests or burrows through the sudden midday darkness.

No total eclipse can last much longer than seven minutes, because of Earth's persistent turning on its axis and the Moon's unwavering march along its orbit. But totality of the briefest duration affords sufficient reason for scientific expeditions and curious individuals to travel halfway around the world, even if they have seen one or more eclipses before.

At totality, when the Moon is a pool of soot hiding the bright solar sphere, and the sky deepens to a crepuscular blue, the Sun's magnificent corona, normally invisible, flashes into view. Pearl and platinum-coloured streamers of coronal gas surround the vanished Sun like a jagged halo. Long red ribbons of electrified hydrogen leap from behind the black Moon and dance in the shimmering corona. All these rare, incredible sights offer themselves to the naked eye, as totality provides the only safe time to gaze at the omnipotent Sun without fear of requital in blindness.

Moments later, the shadow of the Moon passes and the natural world order is restored by the ordinary grace of the Sun's familiar light. But visions of the eclipse persist among viewers, as though a miracle had been witnessed. Is it an accident that the Solar System's lone inhabited planet possesses the only satellite precisely sized to create the spectacle of a total solar eclipse? Or is this startling manifestation of the Sun's hidden splendour part of a divine design?

3

MYTHOLOGY

The planets speak an ancient dialect of myth. Their names recall all that happened before history, before science, when Prometheus hung shackled to that cliff in the Caucasus for stealing fire from the sky, and Europe was not yet a continent but still a girl, beloved by a god, who beguiled her disguised as a bull.

In those days Hermes – or Mercury, as the Romans renamed the Greek messenger god – flew fleet as thought on divine errands that earned him more mentions in the annals of mythology than any other Olympian: after the goddess of the harvest lost her only daughter to the god of the underworld, Mercury was sent to negotiate the victim's rescue, and drove her home in a golden cart pulled by black horses. When Cupid got his wish, making Psyche immortal and therefore fit to marry him, it was Mercury who led the bride into the palace of the gods.

The planet Mercury appeared to the ancients, as it appears to the naked eye today, only on the horizon, where it coursed the twilight limbo between day and night. Swift Mercury either heralded the Sun at dawn, or chased after it through dusk. Other planets – Mars, Jupiter, Saturn – could be seen shining high in the sky all night for months on end. But Mercury always fled the darkness for the light, or vice versa, and hastened from view within an hour's time. Likewise the god Mercury served as a go-between, traversing the realms of the living and the dead, conducting the souls of the deceased down to their final abode in Hades.

Myth may have conferred the god's name on the planet, because it mirrored his attributes, or perhaps the observed behaviour of the planet gave rise to legends of the god. Either way, the union of planet Mercury with divine Mercury – and with Hermes, and the Babylonian deity Nabû the Wise before him – was sealed by the fifth century BC.

The persistent image of Mercury, lean and hell-bent as a marathon runner, personifies dispatch. Wings on his sandals urge him on, spurred faster by the wings on his cap, and the magic powers of his winged wand. Although speed tops the panoply of his powers, Mercury also gained fame as a giant-killer (after he slew thousand-eyed Argus) and as the god of music (because he invented the lyre, and his son, Pan, fashioned the shepherd's pipe of reeds), god of commerce and protector of traders (for which he is remembered in words like 'merchant' and 'mercantile'), of cheats and thieves (since he stole herds from his half-brother Apollo on the very first day of his life), of eloquence (having given Pandora the gift of language), as well as of cunning, knowledge, luck, roads, travellers, young men in general, and herdsmen in particular. His snake-entwined wand, the caduceus, has invoked fertility or healing or wisdom over the ages.

Mercury and his fellow travellers called attention to themselves by moving among the fixed stars, which earned them the name '*planetai*', meaning 'wanderers' in Greek. The orderliness of their motions brought 'cosmos' out of 'chaos' in the same language, and inspired an entire lexicon for describing planetary positions. Just as the gods' names still cling to the planets, Greek terms such as 'apogee', 'perigee', 'eccentricity' and 'ephemeris' endure in astronomical discussions. The first observers to coin such words fill a roster of ancient heroes, from Thales of Miletus (624–546 BC), the founding Greek scientist who predicted a solar eclipse and questioned the substance of the universe, to Plato (427–347 BC), who envisioned the planets mounted on seven spheres of invisible crystal, nested one within the other, spinning inside the eighth sphere of the fixed stars, all centred on the solid Earth.* Aristotle (384–322 BC)

---

* The ancients recognized seven planets: Sun, Moon, Mercury, Venus, Mars, Jupiter and Saturn.

Many worldviews unite in this eighteenth-century sense of the sky.
At centre, the dominant Copernican system includes the four
moons for Jupiter and five for Saturn that had been discovered by
1684, ringed in turn by the zodiac constellations. Ptolemy's Earth-

centred plan stands at upper left. The figures at lower left and upper right show Tycho's compromise, in which the Earth remains central and motionless within the orbit of the Moon, while the Sun goes around them, with all the other planets circling the Sun.

later raised the number of celestial spheres to fifty-four, the better to account for the planets' observed deviations from circular paths, and by the time Ptolemy codified astronomy in the second century AD, the major spheres had been augmented further by ingenious smaller circles, called 'epicycles' and 'deferents', required to offset the admitted complexities of planetary motion.

'I know that I am mortal by nature, and ephemeral,' says an epigraph opening Ptolemy's great astronomical treatise, the *Almagest*, 'but when I trace at my pleasure the windings to and fro of the heavenly bodies I no longer touch earth with my feet: I stand in the presence of Zeus himself and take my fill of ambrosia, food of the gods.'

In Ptolemy's model, Mercury orbited the stationary Earth just beyond the sphere of the Moon. The impetus for motion came from a divine force exterior to the network of spheres. More than a millennium later, however, when Copernicus rearranged the planets in 1543, he argued that the mighty Sun, 'as though seated on a royal throne', actually 'governs the family of planets'. Without specifying the force by which the Sun ruled, Copernicus ringed the planets round it in order of their speed, and set Mercury closest to the Sun's hearth because it travelled the fastest.

Indeed, Mercury's proximity to the Sun dominates every condition of the planet's existence – not just its tantivy progress through space, which is all that can be easily gleaned from Earth, but also its internal conflict, its heat, heaviness, and the catastrophic history that left it so small (only one-third Earth's width).

The pull of the nearby Sun rushes Mercury around its orbit at an average velocity of thirty miles per second. At that rate, almost double the Earth's pace, Mercury takes only eighty-eight Earth-days to complete its orbital journey. The same Procrustean gravity that accelerates Mercury's revolution, however, brakes the planet's rotation about its own axis. Because the planet forges ahead so much faster than it spins, any given locale waits half a Mercurian year (about six Earth-weeks) after sunrise for the full light of high noon. Dusk finally descends at year's end. And once the long night commences, another Mercurian year must pass before the Sun rises again. Thus the years hurry by, while the days drag on for ever.

Mercury most likely spun more rapidly on its axis when the Solar System was young. Then each of its days might have numbered as few as eight hours, and even a quick Mercurian year could have contained hundreds such. But tides raised by the Sun in the planet's molten middle gradually damped Mercury's rotation down to its present slow gait.

Day breaks over Mercury in a white heat. The planet has no mitigating atmosphere to bend early morning's light into the rosy-fingered dawn of Homer's song. The nearby Sun lurches into the black sky and looms enormous there, nearly triple the diameter of the familiar orb we see from Earth. Absent any aegis of air to spread out and hold in solar heat, some regions of Mercury get hot enough to melt metals in daylight, then chill to hundreds of degrees below freezing at night. Although the planet Venus actually grows hotter overall, because of its thick blanket of atmospheric gases, and Pluto stays altogether colder on account of its distance from the Sun, no greater extremes of temperature coexist anywhere in the Solar System.

The drastic contrasts between day and night make up for the lack of seasonal changes on Mercury. The planet experiences no real seasons, since it stands erect instead of leaning on a tilted axis the way Earth does. Light and heat always hit Mercury's equator dead on, while the north and south poles, which receive no direct sunlight, remain relatively frigid at all times. In fact, the polar regions probably harbour reservoirs of ice inside craters, where water delivered by comets has been preserved in perpetual shadow.

Mercury usually eludes observation from Earth by hiding in the Sun's glare. The planet becomes visible to the unaided eye only when its orbit carries it far to the east or west of the Sun in Earth's skies. During such 'elongations', Mercury may hover on the horizon every morning or evening for days or weeks. It remains difficult to see, however, because the sky is relatively bright at those times, and the planet so small and so far away.

Even as Mercury draws closest to Earth, fifty million miles still separate it from us, which is quite remote compared to the Moon's average distance of only a quarter of a million miles. Moreover, the illuminated portion of Mercury thins to a mere crescent as the planet approaches Earth. Only the most diligent observers can spot it, and only with good fortune. Copernicus, caught between the miserable weather in northern Poland and the reclusive nature of Mercury, fared worse than his earliest predecessors. As he grumbled in *De Revolutionibus*, 'The ancients had the advantage of a clearer sky; the Nile – so they say – does not exhale such misty vapours as those we get from the Vistula.'

Copernicus further complained of Mercury, 'The planet has tortured us with its many riddles and with the painstaking labour involved as we explored its wanderings.' When he aligned the planets in the Sun-centred universe of his imagination, he used observations made by other astronomers, both ancient and contemporary. None of those individuals, however, had sighted Mercury often enough or precisely enough to help Copernicus establish its orbit as he had hoped.

The Danish perfectionist Tycho Brahe, born in 1546, just three years after Copernicus's death, amassed a great number of Mercury observations – at least eighty-five – from his astronomical castle on the island of Hven, where he used instruments of his own design to measure the positions of each planet at accurately noted times. Inheriting this trove of information, Brahe's German associate Johannes Kepler determined the correct orbits of all the wanderers in 1609 – 'even Mercury itself.'

It later occurred to Kepler that although Mercury remained hard to see at the horizon, he might catch it high overhead on one of those special occasions, called a 'transit', when the planet must cross directly in front of the Sun. Then, by projecting the Sun's image through a telescope onto a sheet of paper, where he could view it safely, he would track Mercury's dark form as it travelled from one edge of the Sun's disk to the other over a period of several hours. In 1629 Kepler predicted such a 'transit of Mercury' for November 7, 1631, but he died the year before the event took place. Astronomer Pierre Gassendi in Paris, primed by Kepler's prediction, prepared to watch the transit, then erupted into an extended metaphor of mythological allusions when the event unfolded more or less on schedule and he alone witnessed it through intermittent clouds.

'That sly Cyllenius,' wrote Gassendi, calling Mercury a name derived from the Arcadian mountain Cyllene, where the god was born,

> introduced a fog to cover the earth and then appeared sooner and smaller than expected so that he could pass by either undetected or unrecognized. But accustomed to the tricks he played even in his

infancy [i.e., Mercury's early theft of Apollo's herds],
Apollo favoured us and arranged it so that, though
he could escape notice in his approach, he could
not depart utterly undetected. It was permitted me
to restrain a bit his winged sandals even as they
fled. [. . .] I am more fortunate than so many of
those Hermes-watchers who looked for the transit
in vain, and I saw him where no one else has seen
him so far, as it were, 'in Phoebus' throne, glittering
with brilliant emeralds.'*

Gassendi's surprise at Mercury's early arrival – around
9 a.m., compared to the published prediction of midday
– cast no aspersions on Kepler, who had cautiously
advised astronomers to begin searching for the transit
the day before, on November 6, in case he had erred in
his calculations, and by the same token to continue
their vigil on the 8th if nothing happened on the 7th.
Gassendi's comment about the small size of Mercury,
however, generated big surprise. His formal report
stressed his astonishment at the planet's smallness,
explaining how he at first dismissed the black dot as a
sunspot, but presently realized it was moving far too
quickly to be anything but the winged messenger himself. Gassendi had expected Mercury's diameter to be
one-fifteenth that of the Sun, as estimated by Ptolemy

* Gassendi quotes here from Ovid, referring to the Sun god Apollo
by his other name, Phoebus.

fifteen hundred years before. Instead, the transit revealed Mercury to be only a fraction of that dimension
– less than one-hundredth the Sun's apparent width.
The aid of the telescope, coupled with Gassendi's sighting Mercury silhouetted against the Sun, had stripped
the planet of the blurred, aggrandizing glow it typically
wore on the horizon.

Over the next several decades, precise measuring
devices mounted on improved telescopes helped astronomers pare Mercury close to its acknowledged current
size of 3,050 miles across, or less than one three-
hundredth the actual diameter of the Sun.

By the end of the seventeenth century, mystic and
magnetic attractions among the Sun and planets had
been replaced with the force of gravity, introduced by Sir
Isaac Newton in 1687 in his book *Principia Mathematica*.
Newton's calculus and the universal law of gravitation
seemed to give astronomers control over the very
heavens. The position of any celestial body could now
be computed correctly for any hour of any day, and if
observed motions differed from predicted motions,
then the heavens might be coerced to yield up a new
planet to account for the discrepancy. This is how
Neptune came to be 'discovered' with paper and pencil
in 1845, a full year before anyone located the distant
body through a telescope.

The same astronomer who successfully predicted
Neptune's presence at the outer margin of the Solar

System later turned his attention inward to Mercury. In September of 1859, Urbain J. J. Leverrier of the Paris Observatory announced with some alarm that the perihelion point of Mercury's orbit was shifting ever so slightly over time, instead of recurring at the same point in each orbit, as Newtonian mechanics required. Leverrier suspected the cause to be the pull of another planet, or even a swarm of small bodies, interposed between Mercury and the Sun. Returning to mythology for an appropriate name, Leverrier called his unseen world Vulcan, after the god of fire and the forge.

Although the immortal Vulcan had been born lame and ever walked with a limp, Leverrier insisted his Vulcan would hasten around its orbit at quadruple Mercury's speed, and transit the Sun at least twice a year. But all attempts to observe those predicted transits failed.

Astronomers next sought Vulcan in the darkened daytime skies around the Sun during the total solar eclipse of July 1860, and again at the August 1869 eclipse. Enough scepticism had developed by then, after ten fruitless years of hunting, to make astronomer Christian Peters in America scoff, 'I will not bother to search for Leverrier's mythical birds.'

'Mercury was the god of thieves,' quipped French observer Camille Flammarion. 'His companion steals away like an anonymous assassin.' Nevertheless the quest for Vulcan continued through the turn of the

45

century, and some astronomers were still pondering the whereabouts of Vulcan in 1915, the year Albert Einstein told the Prussian Academy of Sciences that Newton's mechanics would break down where gravity exerted its greatest power. In the Sun's immediate vicinity, Einstein explained, space itself was warped by an intense gravitational field, and every time Mercury ventured there, it sped up more than Newton had allowed.

'Can you imagine my joy,' Einstein asked a colleague in a letter, 'that the equations of the perihelion movement of Mercury prove correct? I was speechless for several days with excitement.'

Vulcan fell from the sky like Icarus in the wake of Einstein's pronouncements, while Mercury gained new fame from the role it had played in furthering cosmic understanding.

Still Mercury frustrated observers who wanted to know what it looked like. One German astronomer postulated a dense cloud layer completely shrouding Mercury's surface. In Italy, Giovanni Schiaparelli of Milan decided to track the planet overhead in daylight, despite the Sun's glare, in the hope of getting clearer views of its surface. By pointing his telescope upward into the midday sky, instead of horizontally during dawn or dusk, Schiaparelli avoided the turbulent air on Earth's horizon, and also succeeded in keeping Mercury in his sights for hours at a time. Beginning in 1881, avoiding coffee and whisky lest they dull his vision, and forswearing tobacco to the same end, he observed the planet on high at its every elongation. But the pallor of Mercury against the daytime sky confounded his efforts to perceive surface features. After eight years at this Herculean task, Schiaparelli could report nothing but 'extremely faint streaks, which can be made out only with greatest effort and attention'. He sketched these streaks, including one that took the shape of the number five, on a rough map of Mercury he issued in 1889.

A more detailed map followed in 1934, drawn as the culmination of a decade-long study by Eugène Antoniadi at the Meudon Observatory outside Paris. By his own admission, Antoniadi saw little more than Schiaparelli, but, being an excellent draughtsman and having a bigger telescope, he rendered his faint markings with better shading, and named them for Mercury's classical associations: Cyllene (for the god's natal mountain), Apollonia (for his half-brother), Caduceata (for his magic wand), and Solitudo Hermae Trismegisti – the Wilderness of Thrice-Great Hermes. Although these suggestions have disappeared from modern maps, two prominent ridges discovered on Mercury by space-craft imaging are now named 'Schiaparelli' and 'Antoniadi'.

Both Schiaparelli and Antoniadi assumed, given the persistence of the features they discerned over long hours of observation, that only one side of Mercury ever came into view. They thought the Sun had locked the

47

little planet into a pattern that flooded one of its hemispheres with heat and light while leaving the other in permanent darkness. Likewise many of their contemporaries and most of their followers up to the mid-1960s believed that Mercury maintained eternal 'day' on one side and 'night' on the other. But the Sun constrains the rotation and revolution of Mercury according to a different formula: the planet spins around its axis once every 58.6 days – a rate rhythmically related to its orbital period, so that Mercury completes three turns on its axis for every two journeys around the Sun.

The 3:2 pattern affects observers on Earth by repeatedly offering them the same side of Mercury six or seven apparitions in a row. Schiaparelli and Antoniadi indeed beheld an unchanging face of Mercury throughout their studies, and must be forgiven for reaching the wrong conclusion about its rotation, since the planet's behaviour indulged them in their error.

Throughout the twentieth and into the twenty-first century, Mercury has continued to be a difficult target. Even the Hubble Space Telescope, orbiting above the Earth's atmosphere, avoided looking at Mercury, for fear of pointing its delicate optics so dangerously close to the Sun, and only one spacecraft has so far braved the hostile heat and radiation of the near-Mercury environment.

Mariner 10, Earth's emissary to Mercury, flew by the planet twice in 1974 and once more in 1975. It relayed thousands of pictures and measurements of a landscape riddled with crater holes, from small bowls to giant basins. Light or dark trails of debris marked the places where newer assaults had overturned the rubble of the old. Lava that flowed among the impact scars had smoothed over some of the depressions, but overall poor battered Mercury preserved a clear record of the era, ended nearly four billion years ago, when leftover fragments of the Solar System's creation menaced the fledgling planets.

The most violent attack on Mercury inflicted a wound eight hundred miles wide, which has been named Caloris Basin ('the Basin of Heat'). The mile-high mountains on Caloris's rim must have sprung up in response to the massive impact explosion that excavated the Basin, and all around the mountains, further signs of disturbance lay in ridges and rough ground rippling out for hundreds of miles. The collision at Caloris also sent shock waves clear through Mercury's dense, metallic interior, to set off quakes that lifted the crust on the far side of the planet and cut it to pieces.

Mariner 10 photo montages, which captured less than half of Mercury's surface, revealed a network of scarps and fault lines that indicate the whole planet must have shrunk to its present dimensions from some larger beginning. When Mercury's interior contracted, the global crust readjusted itself to fit the suddenly smaller world – like some furtive trick of the god Mercury, disguising himself.

After a thirty-year hiatus in exploration, a new mission called *MESSENGER* (an acronym for MErcury Surface, Space ENvironment, GEochemistry and Ranging) is now en route to Mercury. Launched in August 2004, but unable to fly as quickly or directly as its namesake, the craft will not reach Mercury's vicinity until January 2008. At first sight of the planet, *MESSENGER* will start a detailed mapping effort requiring three flybys of Mercury over the following three years, while the spacecraft orbits the Sun, protected under a sunshade made of ceramic cloth. Then, in March 2011, *MESSENGER* will manoeuvre into orbit around Mercury itself, for a year-long odyssey (as measured in Earth time) to monitor the planet through two of its long days. Circling Mercury rapidly and repeatedly every twelve hours, *MESSENGER* will function as a new oracle, streaming answers to the questions posed by anxious truth-seekers on Earth.

4

BEAUTY

For a breeze of morning moves,
And the planet of Love is on high,
Beginning to faint in the light that she loves
On a bed of daffodil sky,
To faint in the light of the sun she loves,
To faint in his light, and to die.

Alfred, Lord Tennyson, 'Maud'

Now 'morning star', now 'evening star', the bright ornament of the planet Venus plays prelude to the rising Sun, or postscript to the sunset.

For months at a time Venus will vault the eastern horizon before dawn and linger there through daybreak, the last of night's beacons to fade. She begins these morning apparitions close to the Sun in time and space, so that she arrives in a lightening sky. But as the days and nights go by, she comes up sooner and ventures further from the Sun, rising while Dawn is still a distant idea. At length she reaches the end of her tether, and the Sun calls her back, making her rise a little bit later each night, till she again verges on the day. Then Venus vanishes altogether for the time it takes her to pass behind the Sun.

After fifty days, on average, she reappears at the Sun's other hand, in the evening sky, to be hailed as evening star for months to come. Shimmering into view as the Sun goes down, Venus hangs alone in the twilight. The first few sunsets find her bathed in the afterglow colours of the western horizon, but at length Venus comes to light already high overhead, where she dominates night's onset. Who knows how many childhood wishes are squandered on that planet before the gathering darkness brings out the stars?

> Thou fair-haired angel of the evening,
> Now, whilst the sun rests on the mountains, light
> Thy bright torch of love; thy radiant crown
> Put on, and smile upon our evening bed!
> Smile on our loves, and, while thou drawest the
> Blue curtains of the sky, scatter thy silver dew
> On every flower that shuts its sweet eyes
> In timely sleep. Let thy west wind sleep on
> The lake; speak silence with thy glimmering eyes,
> And wash the dusk with silver.
>
> William Blake, 'To The Evening Star'

Hours into the night, Venus still outshines every other light, unless the Moon intrudes to best her. The Moon appears bigger and brighter, by virtue of lying about one hundred times closer to us, though Venus is the larger and the fairer by far. Venus's shroud of yellow-white cloud reflects light much more effectively than the dun-coloured, dust-covered surface of the Moon. Virtually 80 per cent of the Sunlight lavished on Venus just skitters off her cloud tops and spills back into space, while the Moon beams back a mere 8 per cent.

The remarkable brightness of Venus gains lustre from her nearness to Earth. Venus comes within twenty-four million miles of Earth at closest approach – closer than any other planet. (Mars, Earth's second nearest neighbour, always stays at least thirty-five million miles away.) Even when Venus and Earth recede as far from each other as poss-ible, separated by more than one hundred and fifty million miles, Venus retains her superlative brilliance for Earth-bound observers. On the scale of 'apparent magnitude' astronomers use to compare the relative brightness of heavenly bodies, Venus far exceeds the most luminous stars.*

> What strong allurement draws, what spirit guides,
> Thee, Vesper! brightening still, as if the nearer
> Thou com'st to man's abode the spot grew dearer
> Night after night?
>
> William Wordsworth, 'To the Planet Venus'

---

* The faintest stars visible to the naked eye are those of the sixth magnitude. First-magnitude stars are one hundred times brighter, and the very brightest stars rank at zero, or even −1. Bright Venus reaches −4.6, the full Moon −12, and the Sun −27.

Galileo Galilei (1564–1642), whose telescope enabled him to discover the phases of Venus, depicted them this way.

The nearer Venus draws to Earth, the brighter she appears, naturally enough. Yet as her glow crescendos, the globe of Venus actually diminishes from full to gibbous through quarter and then crescent phase. Like the Moon, Venus appears to change shape as she moves along her orbit, and by the time she reaches her closest, most vivid aspect in our skies, only about one-sixth of her visible disk remains illuminated. But proximity stretches this little sliver to a great length, allowing the perceived brightness of Venus to increase even as she thins and wanes away.

Watching Venus through a telescope or binoculars every evening over a period of months shows how she gains in height and brightness as her disk shrinks, and vice versa. Little else becomes apparent, however, since none of Venus's surface features can ever be discerned by sight through her cloud deck. Thus the very clouds that account for her blatant visibility also act to veil her.

Those who know just where to look can sometimes pick out the steady white light of Venus against the light blue background of a fully daylit sky. Napoleon spotted Venus that way while giving a noon address from the palace balcony at Luxembourg, and interpreted her daytime venue as the promise (later fulfilled) of victory in Italy.

On Moonless nights when Venus is nigh, her strong light throws soft, unexpected shadows onto pale walls or patches of ground. The faint silhouette of a Venus shadow, which evades detection by the colour-sensitive inquiry of a direct gaze, often answers to sidelong glances that favour the black-and-white acuity of peripheral vision. But no matter how avidly you hunt the elusive Venus shadow with eyes averted and downcast, your search may still prove vain, while overhead, as though to mock you, the planet's dazzle mimics the landing beam of an oncoming aeroplane, even triggers police reports of unidentified flying objects.

> I stopped to compliment you on this star
> You get the beauty of from where you are.
> To see it so, the bright and only one
> In sunset light, you'd think it was the sun
> That hadn't sunk the way it should have sunk,
> But right in heaven was slowly being shrunk

57

So small as to be virtually gone,
Yet there to watch the darkness coming on –
Like someone dead permitted to exist
Enough to see if he was greatly missed.
I didn't see the sun set. Did it set?
Will anybody swear that isn't it? . . .

Robert Frost, 'The Literate Farmer and the Planet Venus'

Ancient legends celebrated the beauty of planet Venus by declaring her not only divine but also womanly – perhaps because her visitations generally lasted a significant nine months. Although Venus orbits the Sun in just 224 Earth-days, the Earth's own orbital motions help govern Venus's observed behaviour. As seen from the moving Earth, Venus averages 260 days as either morning star or evening star, coinciding with the human gestation period of 255 to 266 days.

The Chaldeans called the planet Ishtar, the love goddess ascending the heavens, and to the Semitic Sumerians she was Nin-si-anna, 'the Lady of the Defences of Heaven'. Her Persian name, Anahita, associated her with fruitfulness. The dual (dawn and dusk) nature of Venus cast her by turns as virgin or vamp to her worshippers.

Ishtar metamorphosed into Aphrodite, the Greek incarnation of love and beauty. She became the Venus of the Romans, revered by the historian Pliny for spreading a vital dew to excite the sexuality of earthly creatures. In China, Venus blended male and female genders in a married couple consisting of the husband evening star, Tai-po, and his wife, the morning star, Nu Chien.

Only the Mayas and the Aztecs of Central America seem to have seen Venus as consistently male, the twin brother of the Sun. The rhythmic association between Venus and the Sun inspired meticulous astronomical observations and complex calendar reckoning in those cultures, as well as blood rituals to recognize the planet's descent into the underworld and subsequent resurrection.

In North America, among the Skidi Pawnee, the veneration of Venus involved human sacrifice to ensure her return. The last teenage girl known to have died in such devotions was kidnapped and ceremonially killed on 22 April 1838.

As a symbol of loveliness, Venus figures in three paintings by Vincent Van Gogh. His *Starry Night* of June 1889, the best-known example, depicts Venus as the bright orb low to the east of the village of Saint-Rémy, during the time the artist's dementia confined him to an asylum there. Art historians and astronomers have also definitively identified Venus in *Road with Cypress and Star*, which Van Gogh completed in mid-May 1890, the day before he left Saint-Rémy. A few weeks later, in Auvers-sur-Oise, near Paris, where he created eighty works in the two months before his suicide, Van Gogh depicted Venus for the last time, inside a scintillating halo, hovering above the west chimney of *White House at Night*.

> Venus voyages . . . but my voice falters;
> Rude rime-making wrongs her beauty,
> Whose breasts and brow, and her breath's sweetness
> Bewitch the worlds.
>
> C. S. Lewis, 'The Planets'

If ever two worlds invited comparison, the twin sisters Earth and Venus lay such a claim, for these planets are almost identical in size, and orbit the Sun at similar distances. Early discoveries about Venus from afar – especially the detection of her atmosphere by Russian astronomer and poet Mikhail Lomonosov in 1761 – fanned widespread fantasies of a lush abode of Earth-like life.

Recent research, however, has exposed only the most glaring contrasts between the two planets. Although at an earlier epoch Venus probably possessed many of the same attributes as Earth, including once-abundant seas, her water has all boiled away. Now Venus parches and bakes under an obscuring sky that blocks light but traps heat, and bears down upon her surface with heavy pressure.

The ten Russian *Venera* and *Vega* spacecraft that successfully landed on Venus between 1970 and 1984 barely had time to take a few pictures and measurements, or quickly sample the surroundings, before succumbing to the harsh conditions. Within an hour or so of arrival, each vehicle either melted in the heat or crumpled under atmospheric pressure comparable to that found underwater on Earth, nearly three thousand feet below sea level.

Discoveries of the drastic differences between Earth and Venus evoked surprise sometimes expressed in moral terms, as though one sister had chosen the right course while the other veered down an errant path. Nevertheless Venus, the wayward sister, preaches an important cautionary tale to careless humans, for her hostile environment proves how even small atmospheric effects can conspire over time to convert an earthly paradise into a hellfire cauldron. Indeed, much current study of Venus aims to save humanity from itself by verifying, for example, the destruction that chlorine compounds wreak in high-altitude clouds.

> And art thou, then, a world like ours,
> Flung from the orb that whirled our own
> A molten pebble from its zone?
> How must the burning sands absorb
> The fire-waves of the blazing orb,
> Thy chain so short, thy path so near
> Thy flame-defying creatures hear
> The maelstroms of the photosphere!
>
> Oliver Wendell Holmes, 'The Flâneur'*

Differences between Earth and Venus doubtless began in their youth, with the Sun beating hotter on the closer

* Holmes, a practising physician and professor of anatomy at Harvard, as well as a poet, essayist, novelist and amateur astronomer, wrote this poem after seeing a transit of Venus on 6 December 1882.

of the two sisters. The Sun warmed the waters of Venus until they rose in steam, until water vapour and the hot breath of volcanic eruptions enveloped the planet. These gases then did the work of greenhouse glass: they allowed the Sun's heat to reach the surface of Venus, but refused to let heat escape. Instead of dissipating into space, the heat rebounded back down to ground level and made the surface hundreds of degrees hotter still.

High over Venus, Sunlight split the water vapour into its components, hydrogen and oxygen, and the lighter hydrogen escaped the planet's hold. Oxygen remained behind; it recombined with the surface rocks on Venus, and with gases vented by volcanoes, to create an atmosphere consisting almost entirely (97 per cent) of carbon dioxide, the most efficient and pernicious of all greenhouse gases. Today, although only a trickle of solar energy penetrates Venus's cloud cover and arrives at the surface, the greenhouse effect keeps temperatures above eight hundred degrees Fahrenheit all around the planet, day side and night side, even at the poles. Ice on Venus? Liquid water? Impossible, although traces of water vapour do lace the sky.

The abundant carbon dioxide weighs on Venus's hot terrain with ninety times the pressure of Earth's atmosphere. On and just above the surface, where the Russian robot explorers conducted their brief surveys, the Venusian air is thick but transparent, enabling the spacecraft's cameras to see clear to the horizon in the dim available light. All the light was red. Since only the long red wavelengths of light survive the journey down through the cloud canopy, the landscape presents itself as a monochrome in the sepia tones of old photographs. When night takes even this low-level light away, the vista glows in the dark. Its red-hot rocks, cooked halfway to their melting point by the ambient heat and pressure, resemble the embers of a fire.

Some twenty miles above the surface, the clouds set in, in layers fifteen miles thick, admitting no breaks in their coverage. They bar the Sun from ever showing itself at all during the whole course of the long Venusian day. The planet turns so slowly that a single day takes what would be reckoned as two months on Earth just to get from Sunrise to Sunset. Diffuse signs of the Sun's light spread slowly from horizon to horizon as the hours pass, but even the brightest hours of the day stay as dimly lit as vespertide. At night, no stars or other planets ever appear through the perpetual overcast.

Venusian clouds comprise large and small droplets of real vitriol – sulphuric acid along with caustic compounds of chlorine and fluorine. They precipitate a constant acid rain, called virga, that evaporates in Venus's hot, arid air before it has a chance to strike the ground.

Scientists suspect that every several hundred million years the clouds may be remade by a fresh injection of sulphur from global tectonic upheaval on Venus, but failing that, they probably never part.

At their topmost layer, the Venusian clouds display dark swirls when imaged in ultraviolet light. These markings change rapidly, revealing the high velocity at which the clouds roll by – about 220 miles per hour – circling Venus every four Earth-days on fierce winds. Lower down in the atmosphere the winds slacken gradually until they reach the surface, where they don't so much blow as creep across the planet at two to four miles an hour.

Fast or slow, the winds head ever westerly, the same way Venus turns. In contrast to all the other planets, Venus rotates to the west, even as she revolves eastward with them around the Sun. If you could see the Sun rise on Venus, it would come up in the west and set in the east. Astronomers attribute the backward spin to some violent collision that overturned Venus early in her history. The same presumed impact could explain Venus's very slow rotation rate, or perhaps it is the Sun that impedes the planet's spin by raising tides in the vast ocean of Venusian air.

> Deep within that
> libidinous albedo
> temperatures are hot enough
> to boil lead,
> pressures
> 90 times more unyielding
> than Earth's.
> And though layered cloud-decks

```
                    and haze strata
                seem to breathe
                        like a giant bellows,
            heaving and sighing
                every 4 days,
            the Venerean cocoon
                        is no cheery chrysalis
            brewing a damselfly
                or coaxing life
                        into a reticent grub,
            but a sniffling atmosphere
            40 miles thick
                of sulphuric, hydrochloric,
                and hydrofluoric acids
            all sweating
                        like a global terrarium,
                cutthroat, tart, and self-absorbed.
```

Diane Ackerman, 'Venus'

After hiding for an eternity beneath her seething atmosphere, Venus's surface has surrendered to radar examination by Earth-based telescopes and a series of orbiting spacecraft. The finest of these envoys, *Magellan*, circumnavigated Venus eight times a day for four years beginning in 1990.* *Magellan* resolved the planet's vague face into distinct features, most of which turned

* The spacecraft honours Portuguese explorer Ferdinand Magellan, who planned the first circumnavigation of the Earth, and set out from Spain with five ships in 1519. Although Magellan died en route during a battle in the Philippines, one of his ships and a skeleton crew completed his mission, returning to Spain in 1522.

out to be volcanoes of every variety on plains paved with lava.

Magellan's sudden identification of millions of land forms fomented a crisis in nomenclature. The International Astronomical Union responded with an all-female naming scheme that evoked a goddess or giantess from every heritage and era, along with heroines real or invented. Thus the Venusian highlands, the counterparts to Earth's continents, took the names of love goddesses – Aphrodite Terra, Ishtar Terra, Lada Terra, with hundreds of their hills and dales christened for fertility goddesses and sea goddesses. Large craters commemorate notable women (including American astronomer Maria Mitchell, who photographed the 1882 transit of Venus from the Vassar College Observatory), while small craters bear common first names for girls. Venus's scarps hail seven goddesses of the hearth, small hills the goddesses of the sea, ridges the goddesses of the sky, and so on across low plains named from myth and legend for the likes of Helen and Guinevere, down canyons called after Moon goddesses and huntresses.

The only male name on the map of Venus – the great mountain range Maxwell Montes – belongs to Scottish physicist James Clerk Maxwell, who performed pioneering work on electromagnetic radiation during the nineteenth century. When the five-mile-high peaks were detected in the 1960s via Earth-based radar studies

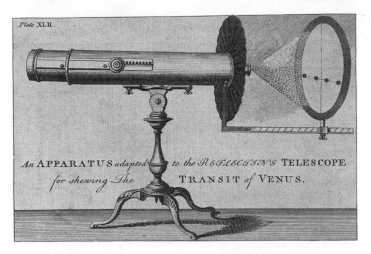

Plate XLII.

An APPARATUS *adapted* to the *REFLECTING* TELESCOPE *for shewing The* TRANSIT *of* VENUS.

Jeremiah Horrocks (1618–1641) became the first person ever to see a transit of Venus on 4 December 1639, observing the event through a small telescope from his home in Hoole, north of Liverpool, and measuring the diameter of Venus as she crossed the Sun.

made possible by Maxwell's insights, it seemed fitting to attach his name to them. For several decades after discovery, Maxwell Montes stood as the sole eponymous feature on the planet, while the low regions on either side of the mountains were designated simply as Alpha Regio and Beta Regio ('A' region and 'B' region). When *Magellan* arrived thirty years later, and its revelations gave rise to names derived from women's history, no one wished to evict Maxwell from his rightful place on Venus.

> Yes, the faces in the crowd,
> And the wakened echoes, glancing

From the mountain, rocky browed,
And the lights in water dancing –
Each my wandering sense entrancing,
Tells me back my thoughts aloud,
All the joys of Truth enhancing
Crushing all that makes me proud.

James Clerk Maxwell,
'Reflex Musings: Reflections from Various Surfaces'*

Magellan's radar images look like night-time aerial reconnaissance photos, except that instead of providing a visual record, their blacks and whites reflect the varying textures of Venus's exposed beauty: hundreds of thousands of small Venusian volcanoes pop out as bright (rough) bumps against the dark (smooth) background of the plains. On the flanks of giant volcanoes, bright (new) layers of lava drape themselves over the dark (old) flows. Mountainsides glittering in radar brightness seem to boast slopes coated with a veneer of reflective metal, perhaps fool's gold, that adheres to Venusian rock at the cooler temperatures a few thousand feet up.

Etched in these images, Venus reveals her unique oddities, such as overlapping 'pancake dome' volcanoes that rise from surprisingly round bases to flat or softly mounded tops, and her numerous 'coronae', or sets of concentric rings that ornately surround so many of her

* The physicist wrote poems as a hobby, and saw forty-three of them published.

domes, depressions and crowds of small volcanoes. Rushing streams of lava dug the long riverine channels that wind across her ample plains. On her high plateaux, tectonic folding and faulting have decorated several thousand square miles to look like crazy-tiled floors, now called 'tesserae'. Evocative patterns in Venus's extruded lava and cracked ground that reminded scientists of sea anemones and spider webs have become 'anemone volcanoes' and 'arachnoids'.

After amassing their gallery of radar portraits, Venus specialists enhanced many of the images with colour for improved resolution. They chose a fire-and-brimstone palette, beginning with the russet hue of the first photos taken by the Russian *Venera* spacecraft, continuing the theme in ochre, umber, sienna, copper, pumpkin and gold. The vibrant colours suit the seared scenery, the rock that spewed forth as lava and still retains its near-plastic consistency, the massifs ascending to altitude without ever hardening harder than toffee. Bright shades befit the youthful visage of a planet that only recently (within the last half billion years) repaved itself in veritable floods of lava, which welled up and covered over almost every vestige (about 85 per cent) of her ancient past.

Relatively few craters mar the new face of Venus, since the rate of cratering over these past 500,000 years is much reduced from the Solar System's earliest days. Many small would-be intruders are vaporized on their way through the thick atmosphere, never to touch down, so that only the very largest impactors reach the surface intact. These collisions eject copious debris, yet all the rubble hugs close around the crater margins in neat festoons, as though contained there by the heavy air. The atmosphere likewise may have soothed the fury of Venusian volcanoes, compelling their expelled lava to seep and pour rather than erupt with explosive force.

Although *Magellan* witnessed no outflows over the years it observed Venus, some of her volcanoes may well be active. Right now, sulphurous gases hissing from Venusian fumaroles could be making their way up to the clouds above the planet, to augment them and sustain them, and thereby ensure the enduring brightness of Venus to our eyes. That fair appearance of unassailable purity once made Venus the darling of poets, whose words still best express her effect on the night's blue velvet – 'a joy for ever', as Keats said, 'a cheering light / Unto our souls'. But new odes to Venus inspired by informed impressions of her savage beauty will have to tap sprung rhythms to describe it, and perhaps shun rhyme.

5

GEOGRAPHY

OR
ON BECOMING A PLANET

To draw the map of the world, begin at the centre of the universe. This is where the astronomer Ptolemy takes up his geography project in the second century. Having already compiled his famous astronomy book, the *Almagest*, in AD 150, Ptolemy turns to the problem of arranging the earth's eight thousand known localities in their proper relative positions. He can hardly have attempted mapping the ground before mastering the sky, because he requires the Sun and stars to guide the placement of each earthly feature. Without astronomy, Ptolemy knows there can be no geography.

Ideally, Ptolemy wishes to watch which way his shadow falls at noon on certain days of the year in distant capitals, see what constellations appear there at night from one season to the next, and note whether the planets pass directly overhead, or ascend only partway up the sky. Alas, he cannot venture so far. Although the celestial spheres routinely rotate the Sun, Moon, planets and a thousand stars into his view, the ends of the earth elude him.

Rooted at his map table in Alexandria, Ptolemy explores the world through the works of previous – often careless – cartographers and a babble of travellers' reports. Thus he hears the distance from Libya to the country 'where the rhinoceros congregate' variously described by Roman army officers as a forced march of three – or four – months' duration, with no reference to the number of days spent resting en route or even the precise direction taken.

If only those favoured with travel opportunities, laments Ptolemy in *Geographia*, his how-to book for mapmakers, would heed the astronomical landmarks! Lunar eclipses, he says, which may occur as often as once every six months, provide the means for anchoring whole strings of locations east or west of each other at a stroke. Unfortunately, as Ptolemy notes, this potential boon to cartography has gone untapped for the past five hundred years – since the lunar eclipse of 20 September in 331 BC, when Alexander the Great met Darius of Persia on the battlefield. Observers sighted that memorable eclipse over Carthage at the second hour of the evening, and farther east in the Assyrian metropolis of Arbela at the fifth hour, from which facts Ptolemy (correctly) establishes the distance between the two cities as 45 degrees of longitude.*

To gauge latitudes north or south of the equator, Ptolemy counts the stars – those that rise and set over a given region at different times over the course of the year, those that neither rise nor set but always appear as darkness falls, and those that never come into view, though they be well known elsewhere. On the Island of Thulē (Shetland Islands), for example, far up at 63 degrees north, where the longest day lasts a full twenty hours, no one sees the mid-summer return of the Dog Star that marks the flooding of the Nile in Egypt.

Ptolemy assumes the world to measure 18,000 miles around. His predecessor Eratosthenes had figured the earth's circumference at a more generous 25,000 miles in 240 BC, by comparing shadow lengths in two cities along the Nile on the day of the summer solstice, but Ptolemy favours the more recent work of Poseidonius, about 100 BC, who observed the stars to shrink the globe.

Ptolemy's *Geographia* offers instructions for creating globes as well as flat map projections. However, the 'known world', as Ptolemy calls it – or 'the inhabited world' or 'the world of our time' – occupies only half a hemisphere. From the Islands of the Blest off the west

---

* Since the Sun appears to circle the sphere of the earth, 360 degrees around, once every 24 hours, Ptolemy calculates each hour's time difference as 360 divided by 24, or 15 degrees of longitude.

In Ptolemy's second-century universe, the earth nestles inside the celestial spheres carrying the planets and fixed stars.

coast of Africa, it stretches eastward all the way through 'India Beyond the Ganges' to 'Sera', where the Silk Road ends, and south from the 'lands of the unknown Skythians', near the Baltic, to the junction of the Blue Nile with the White. Beyond those borders of familiarity, Ptolemy's depiction of lower Africa widens into blank spaces as the continent approaches the Equator, then fans out at the Tropic of Capricorn into a vague undiscovered country, spreading down and across the southern limit of the map, and rising to meet China at the far eastern margin of the Indian Ocean. It is an entirely landlocked world, with all bays and seas surrounded by empires and satrapies, for none of Ptolemy's sources has ventured far enough by vessel to realize the water's true extent.

'In all subjects that have not reached a state of complete knowledge,' Ptolemy says in *Geographia*, 'whether because they are too vast, or because they do not always remain the same, the passage of time always makes far more accurate research possible; and such is the case with world cartography, too.'

The passage of one thousand years changes the shape of the world map from Ptolemy's vision to a circle centred on Jerusalem. Now Heaven imposes a new focus on geography, directing pilgrims and Crusaders to the Holy Land. Although Ptolemy had orientated his globe with North at the top, the new world, as viewed by the Catholic Church, has taken a quarter-turn counterclockwise, leaving East uppermost instead.

This widespread image, the medieval 'mappa mundi', is divided into three unequal parts, one for each of

Noah's sons: Asia fills the top half, while Europe and Africa stand side by side in the bottom. The borders of the three lands suggest a 'T' inscribed inside an 'O', since the lower length of Asia bisects the circle along its diameter, and the Europe–Africa boundary divides the lower hemisphere in half. At the junction of these two brushstrokes sits Jerusalem.

In lieu of places arranged by latitude and longitude, the mappa mundi gives a global overview overlaid with scrambled bits of knowledge concerning this world and the next. The example installed at Hereford Cathedral around the year 1300 locates the Gates of Paradise, the Tower of Babel, the Ark at rest in Armenia, and the place where Lot's wife changed into a pillar of salt. It situates forty mythical and actual animals near their natural habitats and describes them in accompanying legends, including centaur, mermaid, unicorn, 'giant ants' that 'guard sands of gold', and the lynx that 'sees through walls and pisses black stone'. Stranger still are the map's fifty 'monstrous' races of man – the Arimaspi who 'battle Griffins for emeralds', the Blemyae with their mouths and eyes in their chests. Few of these foreigners display Christian or even human virtues, and only the Corcina people of Asia recall Ptolemy's geography lessons of old, for their shadows are said to 'fall north in winter, south in summer', meaning they live in the tropics.

A single hemisphere still suffices to support the whole

world population of the mappa mundi. Around its circumference a great ocean rims the land in view and presumably flows all around the back. The mappa mundi may appear a flat disk rendered on vellum, but it represents a globe. The challenge for Christopher Columbus will not be to convince his critics of the earth's roundness, but rather that it is smaller than they imagine.

Columbus clings to Ptolemy's belief in a world with a girth of merely 18,000 miles, though he knows that Portuguese navigators estimate it to be at least 24,000. Defying them, Columbus wagers he can cross the unknown waters before his crews die of hunger or thirst.

Officially, as Columbus acknowledges in his log, he leads a religious mission, 'In the name of Our Lord Jesus Christ', sent by the 'most Christian, exalted, excellent, and powerful princes', the King and Queen of Spain, 'to the regions of India, to see the Princes there and the peoples and the lands, and to learn of their disposition, and of everything, and the measures which could be taken for their conversion to our Holy Faith'.

Given his prior experience of the sea and his interest in geography, Columbus vows to press his unique situation:

> I propose to make a new chart for navigation, on which I will set down all the sea and lands of the Ocean Sea, in their correct locations and with their

correct bearings. Further, I shall compile a book and
shall map everything by latitude and longitude. And
above all, it is fitting that I forget about sleeping
and devote much attention to navigation in order
to accomplish this.

At the same time, Columbus must quell the fears of
the ninety-odd roustabouts and officers who accom-
pany him on the three ships.

'This day we completely lost sight of land,' he reports
on Sunday, 9 September 1492,

and many men sighed and wept for fear they would
not see it again for a long time. I comforted them
with great promises of lands and riches. To sustain
their hope and dispel their fears of a long voyage,
I decided to reckon fewer leagues than we actually
made. I did this that they might not think themselves
so great a distance from Spain as they really were. For
myself I will keep a confidential accurate reckoning.

When Columbus makes landfall in the Caribbean,
nothing he discovers among the islands dispels his fixed
idea that he has reached India:

'The woods and vegetation are as green as in April in
Andalucía, and the song of the little birds might make a
man wish never to leave here,' he writes on 21 October.

The flocks of parrots that darken the sun and the
large and small birds of so many species are so

different from our own that it is a wonder. In addition, there are trees of a thousand kinds, all with fruit according to their kind, and they all give off a marvellous fragrance. I am the saddest man in the world for not knowing what kinds of things these are because I am very sure that they are valuable. I am bringing a sample of everything I can.

To be sure, Columbus is no naturalist, yet he cites the parrots over and over. The green and purple birds, identified on mappae mundi as the product of India, testify he has indeed arrived somewhere near his intended destination. The 'mainland' the natives describe at a distance of perhaps ten days' journey must be India. The island they call Cuba, he concludes on 27 October, the day before he lands there, is just 'the Indian name for Japan'.

In his own choice of place names, Columbus honours his Saviour and his sovereigns: San Salvador, Santa María de la Concepcíon, Ferdinandina, Isabela. Dubbing his way through the archipelago, he is barred from a full tour of the region by the grounding of one ship and attempted mutiny aboard another.

On his way home to glory, a February storm blows out of the sea with the force of the Devil. Columbus, fearful the water will swallow him before he can declare his discoveries to the Crown, now draws his chart. He secures the parchment in a waxed cloth, seals the cloth inside a barrel, and throws the barrel to the waves.

Should he perish, the finder of his message may inform the Sovereigns 'how Our Lord has given me victory in everything I desired about the Indies'.

Instead, the map disappears in the storm while its maker lives to command three more westward voyages as Admiral of the Ocean Sea and Viceroy of the Indies. Through all these explorations, Columbus never concedes he has found anything but a shortcut to the Orient. Only after his death, as his remains traipse back and forth across the Atlantic yet again for a second and even a third burial, does the magnitude of his discovery divide the globe into the Old World and the New.

Slowly the outline of the far shore takes form. The piece labeled 'Florida' hangs unattached, floating like a shroud above Columbus's Hispaniola (Haiti and the Dominican Republic), until the ends of the Floridian borderlines connect to a larger landmass. 'America' appears for the first time on a wide new world map in 1507. The territory borrows the Christian name of its frequent visitor Amerigo Vespucci, an Italian merchant and navigator. Vespucci has sailed west with both the Portuguese and the Spanish, riding the rivalry between them, boldly proclaiming their scattered claims a bona fide continent distinct from Asia.

At first Vespucci's first name applies only to the southern half of the New World, but it comes to envelop the northern part as well, as explorers from competing countries push ahead to see what lies beyond.

The Spanish throne wins a great victory when Vasco
Núñez de Balboa falls to his knees on a bare summit of
Panama in September 1513, exulting at first sight of the
Pacific Ocean. It takes him several days to descend from
his encampment through the forest to the shore, where
he wades into the water to baptize it. Sword drawn and
shield held high, Balboa shouts the name of Spain over
this sea and every land washed by it, as though he
already knows it bathes half the world.

In 1520, Ferdinand Magellan ventures with five
Spanish ships into the Pacific, and measures its width
in hardship. 'We were three months and twenty days
without getting any kind of fresh food,' Magellan's Italian
navigator, Antonio Pigafetta, writes of the crossing.

> We ate biscuit, which was no longer biscuit, but
> powder of biscuits swarming with worms, for they
> had eaten the good. It stank strongly of the urine
> of rats. We drank yellow water that had been putrid
> for many days. We also ate some ox-hides that
> covered the top of the mainyard to prevent the yard
> from chafing the shrouds, and which had become
> exceedingly hard because of the sun, rain and wind.
> We left the hides in the sea for four or five days,
> and then placed them for a few moments on top of
> the embers, and so ate them; and often we ate saw-
> dust from boards. Rats were sold for one half-ducado
> apiece, and even then we could not get them.

In the heat of this age of exploration, in 1543, a Polish cleric publishes a book that moves the entire world to a new locale. *De Revolutionibus*, by Nicolaus Copernicus, plucks the earth from its stationary post at the hub of the celestial spheres, and sets it spinning around the Sun, between the orbits of Venus and Mars. The strangeness and unpopularity of Copernicus's opinion nearly silence it, but within one hundred years, against all expectation, the Sun takes over the centre of the universe, and our world voyages as a wandering star.

Doesn't this new planet deserve a name? If Champlain can christen his lake and Hudson his bay, why must the newly mobile globe labour under an old, inaccurate term? 'Earth' recalls the ancient division of all ordinary matter into four elements – earth, water, air, fire – and the designation of earth as the heaviest, least heavenly among them. In that scheme, water flowed over earth, air floated above both, and fire rose through air to the threshold of the celestial spheres, where planets and stars embodied a fifth element – quintessence. With world order shifting on maps of the heavens, might not 'the earth' take a proper name from mythology? But already it is too late to dislodge the old name, too late even to change it from 'earth' to 'water', now that seas can be seen to yawn and stretch in all directions.

Mapmakers decorate the blank expanses of ocean with ships, with whales and sea monsters, with puff-

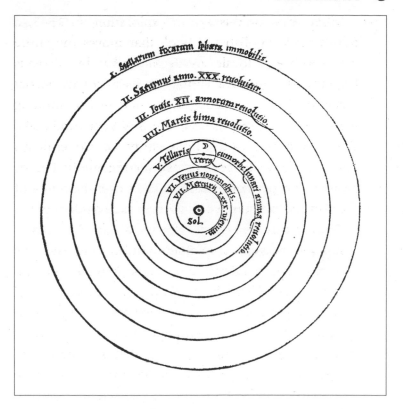

In his book *De Revolutionibus*, Nicolaus Copernicus (1473–1543) transformed the earth into a planet orbiting the Sun.

cheeked cherubs exhaling gales, and also with map titles and legends framed in elaborate cartouches as large as some countries.

At least one compass rose, a flower-like emblem often rendered in gold leaf, indigo and cochineal, now orients each map, with thirty-two painted petals pointing in

85

every possible direction of wind and headway. The rose realizes all the logbook shorthand of exploration's zigzag course – ENE, SSW, NW by N – and mirrors the face of the magnetic compass that dictates those notations.

The magnetic compass, indispensable to mariners since at least the thirteenth century, helps them find the North Star even when clouds obscure it – even when their ship has sailed so far south as to plunge that guiding light below the horizon. Many think the compass needle must be attracted to the Pole Star, if not to some invisible celestial point close by it.

But no, the earth itself is the magnet that draws all compass needles to its iron heart. William Gilbert, an English doctor, discovers this truth through experimentation in 1600, and demonstrates the effect for Queen Elizabeth by using a small spherical magnet to model the earth. Furthermore, Gilbert scorns the universal prohibition against garlic on shipboard, by showing that neither garlic fumes nor garlic smeared on a compass needle can diminish its magnetic power.

The magnetic nature of the earth leads Gilbert and others to suspect magnetism as the force that keeps the planets in their orbits. Newton's universal gravity trumps Gilbert's interplanetary magnetism in 1687, but still the magnetic earth holds promise for navigation. Although compass needles generally tend north, a magnetic compass points slightly east of north in one part of the world, and slightly west of north in another.

Columbus had noticed this shift on his outward voyage, and feared his instrument was failing him. By the seventeenth century, however, cumulative experience suggests the phenomenon may be exploitable. Perhaps the degree of 'variation' of the compass can be measured from place to place, and the featureless oceans resolved into magnetic zones to help sailors establish their whereabouts during weeks or months at sea. This possibility launches the first purely scientific voyage, under the command of Edmond Halley, the only Astronomer Royal ever to win a commission as captain in the Royal Navy.

Between 1698 and 1700, Halley leads two expeditions across the Atlantic Ocean, and also to the Atlantic's northern and southern limits, until stopped by icebergs in fog. Off the coast of Africa and again near Newfoundland, Halley's specially designed flat-bottomed vessel, the *Paramore*, draws friendly fire from English merchantmen and colonial fishermen who mistake her for a pirate ship.

The map Halley publishes in colour in 1701 fills the ocean with curving lines of varying lengths and widths describing degrees of magnetic variation east and west. The continents bordering the Atlantic serve merely to anchor the all-important lines, and to bear the cartouches, whose palm trees, muses and naked natives have been bumped from the busy waters to the empty lands.

Edmond Halley (1656–1742) crossed the Atlantic to test the
earth's magnetic variation, and published this map of his findings
in 1701.

Halley concludes with honesty that magnetic variation will be of no real use to sailors as a means of determining longitude. What's more, he predicts his carefully drawn lines will shift over time as a result of motions deep within the earth. Halley (presciently) envisions the interior of the planet in alternating shells of solid and molten material that control its magnetic behaviour.

Meanwhile Halley's map of magnetic variation, though a disappointment to him and to his fellow seamen, foments a revolution in cartography. Its curved lines connecting points of equal values (to be hailed as Halleyan lines for a hundred years) add a third dimension to printed maps. Other maps of Halley's – of the stars of the southern hemisphere, the Trade Winds, the predicted path of the 1715 solar eclipse – also gain notoriety for their innovations. For his part, he would chart the whole Solar System if only he could gauge the mileage from the earth to the Sun.*

Halley discerns a way to make this key measurement on the special occasion of a transit of Venus: by watching and timing the event from widely separated points on the globe, scientists could triangulate the sky to calculate the distance from the earth to Venus, then deduce

* Kepler's third law of planetary motion, published in 1609, expressed only the relative distances among the planets, based on the revolutionary period of each. No actual distances had yet been calculated.

the earth's distance from the Sun. Halley predicts two transits, for 1761 and 1769, but he will have to live to the age of 105 to see even the first of the pair. For although Venus passes between the Sun and the earth five times every eight years, her tilted orbital path usually carries her above or below the Sun, from our perspective. In order for Venus to be seen crossing the Sun's face, she must also intersect the plane of the earth's orbit – within two days of the earth's intersecting Venus's orbital plane. These stringent requirements permit two transits to follow within eight years of one another, but only a single such pair per century.

'I strongly urge diligent searchers of the heavens (for whom, when I shall have ended my days, these sights are being kept in store) to bear in mind this injunction of mine,' writes Halley in 1716 of the coming Venus transits, 'and to apply themselves actively and with all their might to making the necessary observations.'

When the time of the first transit comes, in June of 1761, Halley's followers face all manner of disasters – hostile armies, monsoons, dysentery, floods, severe cold – to cover prime observing sites in Africa, India, Russia and Canada, as well as in several European cities. Clouds foil most of the expeditions, however, and astronomers' indeterminate results focus even greater attention on the next opportunity, in 1769, which dispatches 151 official observers to 77 locations around the world.

Each group must time the four crucial moments of the transit, called 'contacts', when Venus and the Sun touch rim to rim. The first occurs as Venus appears to attach herself to the outside of the Sun's circle. Second contact soon follows, as Venus enters fully inside the Sun's embrace, but it takes hours for her to achieve third contact on the far side of the solar disk. By fourth contact she has already exited the Sun, and stands on the brink of separation.

Responsibility for the Royal Society's all-important observations at King George III Island (Tahiti) falls to Lieutenant James Cook. He sets out from England the year before, in August of 1768, so as to arrive in time to make preparations that include the building of a secure observatory, Fort Venus.

Saturday 3rd June [1769]. This day prov'd as favourable to our purpose as we could wish, not a Clowd was to be seen the whole day and the Air was perfectly clear, so that we had every advantage we could desire in Observing the whole of the passage of the Planet Venus over the Suns disk: we very distinctly saw an Atmosphere or dusky shade round the body of the Planet which very much disturbed the times of the Contacts particularly the two internal ones. Dr Solander observed as well as Mr Green and my self, and we differ'd from one another in observing the times of the Contacts much more than could

be expected. Mr Greens Telescope and mine were
of the same Mag[n]ifying power but that of the Dr
was greater then ours.

Through no one's fault, astronomers everywhere
encounter the same difficulties as Cook's men in judg-
ing the exact moments of Venus's entry into and exit
from the Sun's disk. The limitations of even the best
available optics undermine everyone's results, and the
international astronomical community must be content
with merely narrowing the earth–Sun distance to something between ninety-two and ninety-six million miles.

Cook turns his attention from Venus to the second,
secret part of his instructions – a sortie through the icy
sea in search of the great southern Terra Incognita. Failing to find it on this quest, he returns home, but mounts
a second discovery attempt in 1772. Through three cold
years of effort Cook, now made Captain, becomes adept
at turning his ship frequently into the wind to shake
the snow from her sails.

Monday 6th February [1775]. We continued to steer
to the South and SE till noon at which time we
were in the Latitude of 58° 15' S Longitude 21° 34'
West and seeing neither land nor signs of any, I
concluded that what we had seen which I named
Sandwich Land was either a group of Isles &c^a or
else a point of the Continent, for I firmly believe
that there is a tract of land near the Pole, which is

the source of most of the ice which is spread over this vast Southern Ocean. [. . .] I mean a land of some considerable extent. [. . .] It is however true that the greatest part of this Southern Continent (supposing there is one) must lay within the Polar Circle where the Sea is so pestered with ice that the land is thereby inaccessible.

Cook's reckoning of latitude and longitude surpasses the accuracy of all who preceded him in such pursuits. By tracking the motion of the Moon against the stars – a method Halley helped to develop – and with the aid of a new timekeeper that keeps up with the master clock back home at the Greenwich Observatory, Cook knows exactly where he is. His maps show others the way from Success Bay in Tierra del Fuego, his source of wood and water, to Botany Bay in Australia, which he named for its abundance of new plant species, and Poverty Bay, New Zealand, where Cook found 'no one thing we wanted'.

Ships laden with surveying instruments – ships that not only cross oceans but can course close to the land all along coastlines and into the mouths of rivers – now start to re-examine the New World with new precision. This is the mission of HMS *Beagle* in 1831, whose captain carries twenty-two of the best available chronometers – timekeepers of the type Cook praised on his second voyage. Bound for a detailed survey of South

America and then home the long way round via the
East Indies, Captain Robert FitzRoy seeks a gentleman
companion who shares his interests in geology and
natural history, and who will pay his own way. Charles
Darwin, a twenty-two-year-old recent college graduate
unsure of his life's vocation, signs on.

The *Beagle* tortures Darwin with seasickness. Although he may freely, legally abandon ship at any port,
he stays the full tour of duty, which lasts five years. He
copes by spending as much time as possible engaged
ashore while FitzRoy coasts the whole of Argentina,
Chile and the Falkland and Galápagos Islands to make
maps.

> 'I stayed ten weeks at Maldonado, in which time a
> nearly perfect collection of the animals, birds and
> reptiles, was procured,'

Darwin reports of the summer of 1832.

> 'I will give an account of a little excursion I made
> as far as the river Polanco, which is about 70 miles
> distant, in a northerly direction. I may mention, as
> proof of how cheap everything is in this country,
> that I paid only two dollars a day, or eight shillings,
> for two men, together with a troop of about a dozen
> riding-horses. My companions were well armed
> with pistols and sabres; a precaution which I
> thought rather unnecessary; but the first piece of

news we heard was, that, the day before, a traveller from Monte Video had been found dead on the road, with his throat cut. This happened close to a cross, the record of a former murder.'

Despite the dangers of the local wars, Darwin still prefers the land to the sea:

August 11th [1833] – Mr Harris, an Englishman residing at Patagones, a guide, and five Gauchos, who were proceeding to the army on business, were my companions on the journey. [. . .] Shortly after passing the first spring we came in sight of a famous tree, which the Indians reverence as the altar of Walleechu. [. . .] About two leagues beyond this curious tree we halted for the night: at this instant an unfortunate cow was spied by the lynx-eyed Gauchos. Off they set in chase, and in a few minutes she was dragged in by the lazo, and slaughtered. We here had the four necessaries of life 'en el campo', – pasture for the horses, water (only a muddy puddle), meat, and firewood. The Gauchos were in high spirits at finding all these luxuries; and we soon set to work at the poor cow. This was the first night which I had ever passed under the open sky, with the gear of the *recado* [saddle] for my bed. There is high enjoyment in the independence of the Gaucho life – to be able at any moment to pull up your horse, and say, 'Here we will pass the night.' The deathlike stillness of the

95

plain, the dogs keeping watch, the gypsy-group of Gauchos making their beds round the fire, have left in my mind a strongly marked picture of this first night, which will not soon be forgotten.

There will be time enough, after Darwin returns to England, for him to marry and put the concerns of his children ahead of his own, to wander in circles for years of private thought as the souvenir birdskins and other mementoes from the Galápagos help him divine the secret of life's diversity.

For now, hunting fossils, 'geologizing', climbing the Andes, he ponders the forces that can uplift such massive mountains over ages, or grind them to gravel, or make them tremble.

February 20th [1835] – The day has been memorable [here] in [. . .] Valdivia, for the most severe earthquake experienced by the oldest inhabitant. I happened to be on shore, and was lying down in the wood to rest myself. It came on suddenly, and lasted two minutes; but the time appeared much longer. The rocking of the ground was most sensible. The undulations appeared to my companion and myself to come from due east; whilst others thought they proceeded from south-west; which shows how difficult it is in all cases to perceive the direction of these vibrations. There was no difficulty in standing upright, but the motion made me

almost giddy. It was something like the movement
of a vessel in a little cross ripple.

Indeed, the continents themselves are voyaging. They
ride as passengers aboard great slabs of the earth's crust
in constant motion. In 1912, German geologist Alfred
Wegener explains that the east coast of South America
complements the western edge of Africa because the
two continents are pieces of the same jigsaw puzzle.
Once in a prehistoric era they lay cheek by jowl, part of
a single land mass Wegener calls 'Pangaea' ('All-earth'),
surrounded by the waters of 'Panthalassa' ('All-sea'),
before geological forces pulled them apart.

Today the Old World and the New continue to recede
from each other along a still-widening rift in the mid-
Atlantic, where molten material wells up from inside
Earth and lays down new ocean floor. As the Atlantic
spreads, the Pacific shrinks. Under the restless coasts of
Peru, Chile, Japan and the Philippines, old, cold ocean
floor is plunging back into Earth's infernal interior, to
the accompaniment of earthquakes and volcanoes, and
sometimes catastrophic tsunamis.

The ocean bottom undergoes constant recycling, and
no part of it is older than two hundred million years.
The continents, in contrast, stay topside through the
ages, eroded but still intact after four billion years.
Instead of sinking under each other, the continents
wrinkle when the stress of contact deforms their crust:

the Appalachian Mountains testify to an ancient collision between Africa and North America, while the ongoing pressure on the Himalayas continues, even now, to increase their altitude.

Modern explorations conducted by submarine and spacecraft reveal the true, apolitical network of Earth's borderlines, hidden underwater. Mid-ocean ridges and complementary coastal trenches divide the surface of the globe into a mosaic of some thirty plates, each one carrying a piece of a continent, a part of a sea floor. The mosaic pattern changes as plates separate, collide, or grind sideways past one another, impelled by the pent-up residual heat of Earth's violent birth and ongoing radioactive decay.

The seismic shocks that pierce Earth during earthquakes permit the deepest possible introspection. They suggest the continents and ocean floors cast only a thin skin, or crust, around the planet. This crust slims to a slender mile under some ocean areas, while the continental crust averages a thickness of twenty miles plus, yet the crust in its entirety accounts for only one-half of one per cent of Earth's mass. The great bulk of the planet (about two-thirds of its mass) consists of the rocky yet fluid mantle roiling between the crust and the core. At the centre of the Earth, part of the iron-nickel core has already cooled to a solid ball. Seismologists can hear it rotating independently inside the still-molten outer core, turning almost one second a day faster than the rest of the world.

Like the hidden levels of the inner Earth, the invisible layers of Earth's atmosphere have also been charted, from low in the troposphere, up through the stratosphere and mesosphere to the top of the thermosphere. The magnetic field and radiation belts surrounding Earth can be mapped from space. Also from space, a network of global positioning satellites can pinpoint locations – even individuals – on the planet with centimetre accuracy, while laser beam reflectors planted on the Moon by Apollo astronauts gauge the exact Earth–Moon distance.

Earth's place in space is now known to such confident extremes of accuracy that the most recent transit of Venus, on 8 June 2004, was relegated to the status of a tourist attraction – a chance to see an anomaly unknown to any living soul, given the date of the previous transit, on 6 December 1882. In the interim between that transit and this, the extent of the known world had expanded to include additional planets of the Solar System, extrasolar planets in the Galaxy, and the configuration of the Milky Way itself, twirling through space with billions of stars in its spiral arms. A longer view into the infinite takes in the other galaxies of our Local Group, and the clusters and superclusters of galaxies stretching out in space and back in time to the birth of the universe. But even this sophisticated sense of our surroundings, like Ptolemy's map, captures only the present moment's self-awareness.

6

LUNACY

During the glory days of the Apollo project, a young astronomer who analysed Moon rocks at a university laboratory fell in love with my friend Carolyn, and risked his job and the national security to give her a quantum of Moon dust.

'Where is it? Let me see!' I demanded at this news. But she answered quietly, 'I ate it.' After a pause she added, 'There was so little.' As though that explained everything.

I was furious. In an instant I had dropped from the giddy height of discovering the Moon right there in Carolyn's apartment to realizing she had eaten it all without leaving a crumb for me.

In a reverie I saw the Moon dust caress Carolyn's lips like a lover's kiss. As it entered her mouth, it ignited on contact with her saliva to shoot sparks that lodged in her every cell. Crystalline and alien, it illuminated her body's dark recesses like pixie powder, thrumming the senseless tune of a wind chime through her veins. By its sacred presence it changed her very nature: Carolyn, the Moon Goddess. She had mated herself to the Moon somehow via this act of incorporation, and that was what made me so jealous.

Of course I had heard the old wives' tales advising women to open their bedroom curtains and sleep in the Moonlight for heightened fertility or a more regular menstrual cycle, but no folklore described powers to be won from the Moon by eating its dust. Carolyn's deed conjured Space Age magic, undreamed of when her mother and mine were new wives.

I still envy Carolyn her taste of the Moon. In reality I know she is married now to a veterinarian in upstate New York and has three grown children. She doesn't glow in the dark or walk on air. She has long since lost all traces of that Moon morsel, which no doubt passed through her body in the usual way. What could it have contained, anyway, to preoccupy me all these years?

A few grains of titanium and aluminium?

Some helium atoms borne from the Sun on the solar wind?

The shining essence of all that is unattainable?

All of the above, probably, all rendered the more extraordinary for having travelled to her across 240,000 miles of interplanetary space, in the belly of a rocket ship, and hand-delivered as the love token of a handsome man. Lucky, lucky Carolyn.

The Apollo astronauts themselves did not intentionally swallow any Moon dust, though it clung to them,
covered their white boots and space suits with grime,
and so climbed with them back into their lunar modules. The moment they removed their bubble helmets, a
smell of spent gunpowder, or of wet ashes in a fireplace,
assailed them. It was the Moon dust, tamely burning in
the oxygen atmosphere the men had carried from home.
Outside on the airless lunar surface, did the trodden
dust give off any odour of its own? Does a tree falling
in a forest make a sound if no one hears?

The astronauts judged the dusty surface of the Moon a
shade of tan, like beach sand, when they looked at it facing Sunward, but said it turned grey when they turned the
other way – and black when they scooped dust samples
into plastic bags. The unearthly glare of unfiltered Sunlight bedevilled their colour and depth perception, and
that of their photographic film as well. Similarly attuned
to the light of Earth's atmosphere, the film formed its own
interpretation of the new landscape's subtle hues and
stark relief, so that in the end the men's pictures betrayed
their colour memories of walking on the Moon.

The view of the Moon from Earth is no less fooled by
tricks of light. How else could the Moon derive its silvery
gleam from dust and rocks dark as soot? The dusky
markings that draw the face of the Man in the Moon
reflect only 5 to 10 per cent of the Sunlight that falls on
them, and the brighter lunar highlands no more than

12 to 18 per cent, making the Moon overall about as shiny as an asphalt roadway. But the rough-hewn lunar surface, sprinkled with ragged particles of Moon dust, multiplies the myriad planes where light may strike and ricochet. Thus the tan, grey, black dust clothes the Moon in white radiance. And seen against the sombre backdrop of the night sky, the Moon appears whiter still.

Whiteness defines our image of the Moon, except for those occasions when it hangs golden on the horizon, burnished by added thicknesses of air, or dips into Earth's shadow and glows red in total lunar eclipse. No one ever seriously believed the Moon looked green in colour, only that it resembled a green cheese – a whitish, splotchy wheel of new-made curds, not yet ripe for eating. True, the Moon may turn blue after a volcano sullies Earth's atmosphere, or be called blue when it becomes full more than once per calendar month, but the reliable whiteness of the ordinary Moon is what grants the idiomatic Blue Moon its air of rarity.

While white light bouncing off the Moon contains every colour, Moonshine perceived on Earth mischievously bleeds familiar sights of any colour. The full wattage of the full Moon dims in comparison to direct Sunlight, by a factor of 450,000, and so falls just below the retina's threshold for colour vision. Even the brightest Moonlight induces pallor in each face it illuminates, and creates shadows like oubliettes, where all who enter disappear.

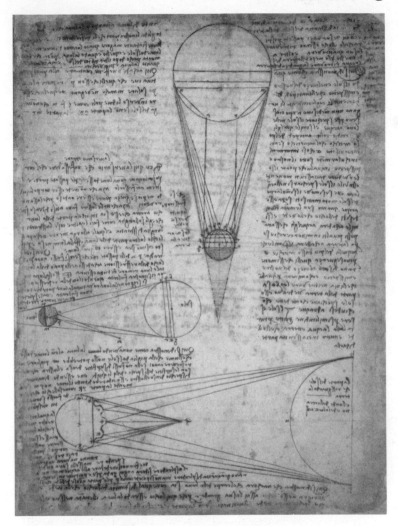

Leonardo da Vinci (1452–1519) pondered the light of the Sun and the Moon and recorded his thoughts between 1506 and 1510 in a notebook, the *Codex Leicester*, named for its longtime owners, the family of the Earl of Leicester.

107

The wan colours of Moonbeams bloom in Moon gardens planted with lilies, angel's trumpet, sweet rocket and the like, all of them white or nearly so, or prized for their nocturnal habits. The giant Moon flower, evening's answer to the morning glory, opens its white petals at day's end, as do its companions the four-o'clocks, the vesper iris, and the night gladiolus. Evening primrose also finds welcome in Moon gardens, despite pink blossoms, because the primrose wafts its perfume after dark.

The Moon itself refuses to be confined to the night. It spends half its time in the daylight sky, where many people take no notice of it at all, or mistake it for a cloud. Only for a few days each month does the Moon truly vanish, rendered invisible in the vicinity of the Sun. The rest of the time the inescapable Moon changes shape by the hour, waxing and waning and whining for attention.

The first sight of the young Moon arrives as a smile at twilight. Though only the slimmest sliver of silver crescent shines on us this early in the Moon's monthly cycle, the rest of the Moon reveals itself in just-discernible form, as though the old Moon were lying in the young Moon's arms. Leonardo da Vinci, sketching the Moon at such a time, recognized the faint light cupped inside the bright crescent as Earthshine. The phantom Moon, Leonardo explained in the crabbed, left-handed mirror writing of his notebooks, catches the Earth's reflection of the Sun, and beams back an attenuated echo.

By the time the Moon moves one quarter of its way around the Earth, the Sun's light covers half the Moon's face, like the icing on a biscuit with precise hemispheres of chocolate and vanilla. Soon the terminator – the day–night line – arches like a bow, and still more of the lunar surface lights up as the Moon gains its gibbous phase. These stages of lunar expansion, unfolding from the dark of the Moon through crescent, quarter, gibbous and full, promise growth. Herbals and farmers' almanacs commend the waxing phases as proper times for sowing peas, harvesting root crops and pruning trees to assure abundant fruit. Timber, however, by the same token, must never be taken in a waxing Moon, because wood wet with rising sap will resist the saw, requiring harder work, and after cutting it will warp.

The full Moon rising at Sunset raises an illusion of grandeur that doubles or triples its apparent size. The splendour of this sight derives from the mind's own sense of the horizon as a faraway place where anything that looms large must be huge indeed. Later on in the night, once the Moon has ascended the sky, where a different distance scale applies, the Moon resumes its normal dimensions, though the world below be mad. Dogs bay, coyotes howl, lycanthropic men morph into werewolves, and vampires prowl under a full Moon. More crimes are committed, more babies are born, more lunatics run amok. Or so some claim, since the full Moon's startling light, almost bright enough to read by, sustains a prevailing expectation of mayhem.

Every full Moon of the year has earned at least one name tying it to lost seasons of tradition – Wolf Moon, Snow Moon, Sap Moon, Crow Moon, Flower Moon, Rose Moon, Thunder Moon, Sturgeon Moon, Harvest Moon, Hunter's Moon, Beaver Moon, Cold Moon – though no such tributes apply to any of the Moon's other phases.

The state of technical fullness, when the Moon stands opposite the Sun in Earth's sky, lasts only a minute in the monthly life of the Moon. A moment later, as the Moon yields to decline, darkness encroaches from the right, retracing the path of the previous light. One by one the features drop from the face of the Man in the Moon – or the rabbit in the Moon, or the toad – in the same order they showed themselves before. First to come or go is high, round Mare Crisium (the Sea of Crises), followed, as in some fantastic Latin incantation, by Lacus Timoris (the Lake of Fear), Mare Tranquillitatis (the Sea of Calm), Sinus Iridum (the Bay of Rainbows), Oceanus Procellarum (the Ocean of Storms), Palus Somni (the Marsh of Sleep).

Nothing could summon water from those dark seas of the Moon because they are, all of them, dry. Nor have the Moon's so-called seas ever known the presence of water. Though the lunar maria hinted of a fluid interconnectedness to the first astronomers who eyed them and named them through telescopes, the first Moon-walkers to tread them retrieved the driest imaginable materials from their shores.

'Bone-dry', the lunar samples were described, though they are much drier than bones, which form inside the Earth's wet living systems, and retain the memory of water long after death.

Dry as dust, then? No, drier still. On Earth, even dust holds water.

Moon rocks set a new standard of dryness, distinguished by the *total* absence of water. Not a drop of water, not a bubble of water vapour lurks in the crystal lattice of any Moon rock among the lunar samples, and no ice ever so much as touched them. Comets, however, have probably tucked odd caches of imported water ice – perhaps ten million tons' worth – in the shadows of unexplored craters near the lunar poles.

Lacking water as a potential ingredient limited the Moon's creativity to a mere one hundred minerals, while the moist Earth has fabricated several thousand mineral varieties. The gems romantically or religiously associated with the Moon – pearl, quartz, opal, moonstone – could never have formed there, for each requires water in one way or another, and the Moon has none to offer.*

---

* Valued as exotica, a single carat of Moon rock sold at auction in 1993 for $442,500. Similarly, a site map of the Moon's Descartes highlands, only slightly used by *Apollo 16* astronauts and bearing smudges of lunar dust, brought $94,000 at a sale in 2001.

The primal lunar scenario currently favoured by planetary scientists explains the Moon's formation and dryness in a single blow: early in the history of the Solar System, a rogue planet on a collision course struck the infant Earth. The impact, thought to have occurred 4.5 billion years ago, melted impactor and impact site alike, and shot hot debris into space. Swarms of dust and rock fragments, lofted into orbit around the stunned Earth, eventually reunited, 4.4 billion years ago, as the Moon. Having been ejected from one common cauldron, Moon rocks chemically resemble Earth rocks, except that they have lost all their water and any other compounds capable of escape as vapours.

The furious pace of lunar assembly generated enough heat to melt the top layers of the new satellite into a global magma ocean, one hundred miles deep. Over time that ocean gradually cooled and hardened to stone. The errant rubble of the Solar System's violent youth, still at large then, bombarded the Moon's smooth new crust, blasting out vast impact basins and craters. Meanwhile radioactive heat trapped inside the young Moon drove more molten rock to the surface, to fill broad basins with black basalt – and paint the Moon's facial features.

The all-encompassing ocean of magma attending the Moon's birth was the first fluid to flow there. The rivers and pools of extruded lava were the last, and those froze up three billion years ago. At that time, the rate of cratering tapered off throughout the Solar System, and the Moon, having expended all its internal heat, solidified through and through, turning into a dry fossil generally considered 'dead' by geological standards.

The parched Moon pulls at Earth's seas as though jealous of them. Twice each day the ocean tides rise and fall to the call of lunar gravity. The waters rise once when they pass beneath the Moon, which makes intuitive sense, but then they rise again after they have been twirled round to the other side of the world, where they face away from the Moon. There you might say they only appear to rise, when really the Earth is being pulled out from under them by the tug of the Moon. Looking at the whole world's waters at once, the ocean directly under the Moon rises in response to the stronger tug of gravity there, while the ocean on Earth's opposite side simultaneously rises as though relieved to feel so little force pulling it in the opposite direction.

Earthly tides answer to solar gravity as well as lunar, but not as much so, for the Sun's greater distance, and its tendency to pull more equitably on all parts of the Earth at once, diminishes its effect on the tides. When the Sun and Moon align with the Earth in a straight line across the heavens, however, as they do at new and at full Moon, then the three bodies conspire to make tides rise higher. Such 'spring tides', which occur in every season, take their name from the rush of waters that

menses with the lapse of the lunar month must be either a coincidence or a mystery.

Even as the Moon draws the oceans to and fro, the Earth drags down the Moon with the superior force of its greater mass. The uneasy power struggle between the bodies has slowed the rotation of the Moon to about ten miles per hour. Spinning this slowly, the Moon takes as long to turn once on its axis as it takes to complete its monthly 1.5-million-mile orbit. The Earth has thus coerced the Moon into a lock-step pattern of rotation and revolution, called 'Earth-lock', that keeps the Moon's same awestruck face trained Earthward at all times. No wonder the Man in the Moon looks so familiar.

Compared to the Moon, the Earth spins like fury, rotating one hundred times faster. Yet the Earth, too, is decelerating, by a few hundredths of a second annually, under the strain of tidal friction. For the Moon's noticeable effect on ocean tides is accompanied by an insidious stretching of the Earth's solid ground. The Moon pulls hardest on whatever part of Earth is nearest, actually raising a bulge. But no sooner has some expanse of Earth's surface risen in response than the Earth's rotation wrenches that region out from under the Moon, and rolls a neighbouring area there instead. With some section of the planet always bulging and then subsiding, the constant friction stays the pace of rotation.

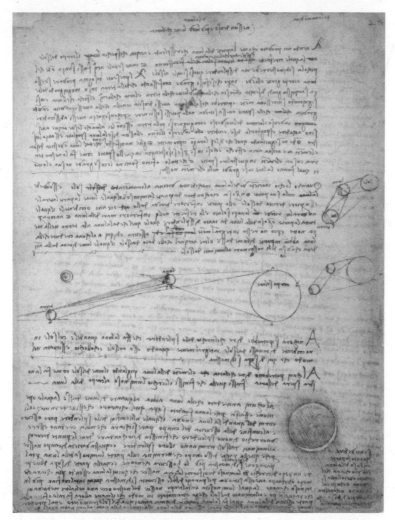

Elsewhere in the *Codex Leicester*, Leonardo considered the ghostly
light contained in the crescent of the young Moon, and recognised
it as the re-reflection of sunlight off the surface of the Earth.

As the Earth slows, the Moon drifts an inch or so further away each year, since the cascade of tidal effects gives a slight boost to the satellite. Eventually the Earth's slowing down and the Moon's slipping away will end in a standoff that stabilizes the Earth's rotation and halts the Moon's retreat. At that point the rotation of both bodies will be synchronized: Earth will eye the Moon with the same wary, one-sided gaze the Moon now fixes on the Earth. Moon worshippers in that distant future will no doubt dwell on the half of Earth where the Moon hovers overhead all the time, while inhabitants of Earth's other hemisphere, 'the far side', will need to journey as much as halfway around the world to get even a glimpse of the Moon.

For now, the almost imperceptible decrease in the Earth's rotation amounts to a mere millisecond every fifty years. But this and other inconstancies have convinced official timekeepers to improve on the Sun, Moon and stars as reliable standards, and occasionally to stitch an extra 'leap second' into the worldwide calendar year. Unlike a leap year, which lasts a day longer than a typical year, a leap second measures the same fraction of time as any other second. But just like the leap year, the leap second sings the frustration of all recorded efforts to base a calendar of human affairs on the motions of the heavenly spheres.*

* A second, which once divided a mean solar day into 86,400 equal parts, is now defined as the time a confined cesium-133 atom takes

The Earth's daily tur[n]
lution around the Sun r[...]
Moon's monthly orbit. Con[...]
cues has ever demanded elab[...]
ing between twelve-month an[...]
(which long ago made the num[...]
or for legislating the duration of the [...]
The mnemonic doggerel of 'Thirty day[s...]
quickly loses its rhyme and metre in th[...]
a requisite number of days into months [...]
with seasons through years to come.

Even though an atomic clock keeps better time than
the dance of the planets, nevertheless it is the clock
that must be readjusted accordingly, and yield to the
authority of the imprecise orbs. For what good is the
smug ability to judge the Earth a second short in her
timing if spring comes when it will?

On the Moon, a single time span – our lunar month
– serves for day and year alike. Over the course of this
daily year, as the Moon turns on its axis and around the
Earth, Sunlight and warmth spread first over one lunar
hemisphere and then the other, granting each about
two weeks of continuous daylight, to be followed by
the frigid two-week night.

Many think of the far side of the Moon as the dark

---

to complete 9,192,631,770 natural vibrations. Since 1972, the International Earth Rotation Service has added 24 leap seconds, always inserted into the first moments of January or July.

of its being perpetually hidden from
, too, goes through phases, which comple-
the fully or partially lit phases we observe on the
near side. Just as the Sun's light bathes half the Earth
all the time, so, too, does it illuminate the sphere of the
Moon.

Apollo astronauts who walked on the Moon landed
on the near side in the early lunar morning, before
the temperature rose to its noon high of 225 degrees
Fahrenheit. Even the last two Apollo crews, who
sojourned on the surface for an elapsed mission time
of three days, came and went within half a morning
on the Moon.

Not one of them set foot on the far side, though they
all saw its strange terrain first-hand from lunar orbit,
and remain the only humans ever to have done so. They
could have exclaimed any expletive or sentiment in the
private thrill of that revelation, since travelling behind
the Moon cut off their radio contact with Houston and
the rest of the world. Apollo command module pilots,
who stayed in orbit while the landing parties worked
the surface, experienced a profound solitude over the
far side, out of touch with all civilization – including
their teammates – for forty-eight minutes out of every
two-hour loop around the Moon. The far side of the
Moon is the one place in the whole Solar System deaf
to Earth's radio noise.

Like the hidden half of any being, the lunar far side

bears scant resemblance to the face the Moon shows the world. More craters abound there, overlapping in profusion, and one sees hardly any of the smooth dark expanses of pooled lava that characterize the near side. The thicker crust on the Moon's back apparently checked the expulsion of lava from within.

All geologic ferment on the Moon ceased about three billennia ago, after the late heavy bombardment cleared the Solar System of most menacing massive projectiles. Today, a ton-mass meteorite strikes the Moon no more than once in three years, on average. The occasional Moonquake can be confidently dismissed as a weak reaction to tidal stress, not the stirrings of a living planet with a liquid core.

Only *micro*-meteorites continue to fall steadily on the dead Moon, thickening the dust by a millionth of a millimetre per year. This influx constitutes the major tectonic force now at work on the Moon. Selenologists call it 'gardening', because the new arrivals mix and turn over the sterile lunar 'soil' as they insert themselves into it. The gentle process barely disturbs the present still life on the Moon – the arrays of scientific instruments, the litter of spent rocket stages, the three parked rover vehicles.

Among the personal talismans intentionally left behind, a posed snapshot of an astronaut and his family calls attention to itself. Someone took care to wrap that photo in plastic for protection – as though anything could happen to it on the Moon's arid, uneventful surface, where a bootprint enjoys a life expectancy of a million years, and every dust particle savours of immortality.

7

SCI-FI

C all me 'It', or call me 'Allan Hills 84001', my given
name – even 'Thing from Mars' will suit. Although
I am only a rock and cannot answer, allow me this
conceit of conscious identity for the space of these few
pages, that I may speak for Mars, whence I travelled via
chance and the laws of physics.

Of the twenty-eight Martian meteorites definitively
identified to date, I am by far the most ancient, and
the only one to show, under microscopic examination,
internal shapes and residues similar to those formed by
primitive terrestrial bacteria. These findings have made
me the most studied rock of all time.

One might surmise I had been contaminated by Earth
life during the thirteen thousand years I lay in the Antarctic ice fields before scientists collected me there in
1984. The scientists certainly assumed contamination,
until they ruled out the possibility, concluding in near

disbelief that it was more likely I had once sheltered small beings on my home planet – creatures perhaps already extinct when an asteroid impact flung me from the Martian surface sixteen million years ago.

My story, consonant with the history of Mars in human regard, seems riveted on Martian life, despite my vagueness on this subject. I have little to say of fossil life, and even less to contribute to conjecture about life on Mars today. Therefore I make no bold claims, lest I be lumped in the company of such fiercely imagined aliens as the giant sand worms of Arrakis, or the wild thoats, green men and great white apes of Barsoom.*

My Martian origin, however, I state as indisputable. My constitution mirrors the chemical makeup of rocks and dust examined *in situ* on the planet's surface and from close orbit by visiting spacecraft. Traces of gases, trapped in glassy bubbles within my matrix, exactly match the sampled atmosphere of Mars, element for element and in the same relative abundance of rare isotopes. My foreign nature could never have been proven before the current age of spacefaring, and yet I came to Earth without benefit of any artificial conveyance.

The collision that launched my journey dug a hole in Mars several miles wide. Astronomers think they have identified that particular crater on satellite images of

* See, for example, Frank Herbert, *Dune* (1965), and Edgar Rice Burroughs, *The Gods of Mars* (1918).

Mars, near a small valley in the southern highlands. The violence of the impact lofted tons of crustal rock into the sparse atmosphere at high speed, and all the fastest-moving pieces – the ones accelerated past the local escape velocity of three miles per second – hastened out of the planet's grip for ever.

As a Martian from a heavily cratered region, I was acquainted with meteorite strikes, and in fact already bore a fracture scar from having been crushed and reheated in a previous impact. But now I found myself *become* a meteorite, or more correctly a meteoroid, that is to say, a true space wanderer loosed from one world and not yet landed at another. Sixteen million years of seemingly aimless rambling at length brought me close enough to Earth to be captured by its gravity, which is three times stronger than the pull of Mars. By long odds I should have disappeared into an ocean, the fate of most meteorites that survive their fiery descent to Earth's surface, but instead I fell near the South Pole, during the last Ice Age, on a bed of frozen water.

Snows came and covered me, folded me into the slowly flowing glacier, and together we crept forward for thousands of years. Not until we reached the Allan Hills, and tried to climb them, did the sharp cliffs and arctic winds pluck me from the ice to leave me lying exposed again.

The scientists arrived on seven snowmobiles, in drag-net formation, hunting dark rocks on the blue-white

127

ice, confident that all such finds would prove extraterrestrial, whether from the Moon, the Asteroid Belt, or Mars. Even though I measure no larger than a squarish softball, or a four-pound potato, they spotted me easily by colour contrast. 'This green rock', I appeared to them in that dazzling expanse of ice and light, only later fading to grey, 'dull grey', in the laboratory.

I was airlifted to the United States, to the Johnson Space Center in Houston, Texas, where my age was established by two independent mother–daughter radio-isotope measurements, one analysing the proportion of samarium that had decayed to neodymium inside me, while the other pursued the radioactive transformation of rubidium to strontium. Both assays yielded the same result, citing a lapse of 4.5 billion years since the time I had crystallized, though these tests said nothing of my provenance. At first examiners took me for an igneous rock from the asteroid Vesta, but later they fired a narrow beam of electrons at some of my grains, exciting my near-surface atoms to emit X-rays that revealed the wider truth of my alien composition, especially the form of iron I contain, which positively identified me as Martian.

My extremely advanced old age sets me apart from other known Martian meteorites. At 4.5 billion years, I quadruple the age of the second-oldest in the group, which suggests that I am a piece of Mars's original planetary crust. No comparable Earth rock has yet been discovered, for the most ancient of these does not exceed 4 billion years, and only a single rock retrieved from the Moon, the so-called 'Genesis Rock', rivals my extra-ordinary antiquity.

A sturdy relic of the Solar System's earliest days, I have maintained myself, virtually unaltered, through aeons likely to have seen me pulverized by impact or melted in a volcano and subsequently reincarnated after cooling.

Mars is a great respecter of longevity. Most of Mars's surface endures today much as it always has, while Earth and Venus go on reinventing theirs through constant upheaval. Yet Mars is no slavish preservationist like the Moon or Mercury, whose static vistas are shaped almost entirely by outside forces. On the contrary, my planet, a globe only half the size of Earth, raised the tallest mountains in the Solar System, carved vast labyrinthine valleys, inundated its lands with liquid water, and then froze to a desert of spectacular dunes in a palette of reds, yellows and browns so vivid as to make Mars, seen from afar, glow like an orange star.

The Martian landscape hosts a desert more dust than sand, and when its fine, smooth, iron-rich particles of rusted dust hang in the sky like a haze of smoke, they share their colour with the air. The pinkish atmosphere, consisting predominantly of carbon dioxide, exerts a barely perceptible pressure at ground level, only one-hundredth that of Earth, but how its winds do spur the

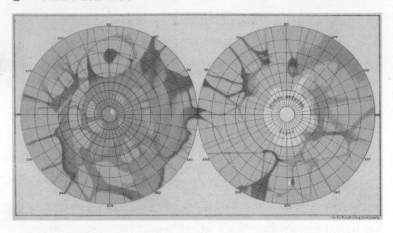

Giovanni Schiaparelli (1835–1910) first noticed *canali* on the surface of Mars during the opposition of 1877, and continued to map and observe the planet for another two decades. Some contemporaries called him 'the Columbus of Mars'.

dust to action! Lone dust devils spiral up and snake across the open spaces. Masses of dust rise in whirling sallow storm fronts fit to rage for days, and sometimes grow into global tempests that envelop the whole of Mars for months, until the dust-laden air finally tires of its burden.

Bright white ice caps at the planet's poles advance and retreat over the ruddy ground in a rhythmic seasonal cycle of changing weather. Between the poles the land divides itself into unequal halves, with most of the ancient, heavily cratered highlands concentrated in the south, where I came from, and the younger, lower-lying plains in the northern hemisphere. Those

far northern plains lie so low that the planet looks lop-
sided, with its south pole four miles further from the
equator than the north pole.

Just north of the equator, mighty Olympus Mons
attained its altitude of Alps atop Rockies upon Himalayas early in Martian history, when the leftover heat of
planetary coalition escaped in eruptions of lava copious
enough to build a dozen monstrous mountains and
scores of smaller ones. Since then Mars's peaks have
received many crater hits on their flanks, but suffered
hardly any erosion. The white water-vapour clouds
around their summits drop no damaging rains down
the mountainsides, and the visiting winds carry only
fine, smooth particles of clay dust, almost too soft to
wear the rock away.

East of Olympus, ancient faults split thousands of
miles of ground asunder to gouge the grand canyons of
the Valles Marineris. Landslides widened these chasms
and rushing waters deepened them, moulding islands
shaped like tear-drops on their floors, but all the planet's
steep-sided valleys stand empty now that the Martian
water supply has disappeared from view.

The warmer, wetter climate of the distant past may
have ended violently, when the impacts that excavated
Mars's widest, deepest basins blew away the water vapour and nitrogen that had once thickened the atmosphere. Then the liquid water fled the surface by every
available means, evaporating and dissipating into space,

131

flowing down into hidden aquifers, hibernating as sub-
terranean permafrost.

My own experience of Mars dates back to a time of
liquid water. Between 1.8 and 3.6 billion years ago, as
narrowly as cosmochemists can estimate, water from
Martian hot springs bathed me, permeated the cracks I
had sustained in earlier shocks, and lined the cracks
with signature veins of carbonate minerals. These mineral deposits now account for one-tenth of my bulk
makeup, and all the signs of life inside me reside within
them.

Startling and unprecedented as my cargo of genuine
extraterrestrials might be, science *embraces* the possibility. Whatever forces sparked the emergence of life on
Earth, three to four billion years ago, could have done
the same on Mars at that same early epoch. Even supposing that only Earth alone among all the planets ever
gave birth to life, it is still conceivable that at least one
archaeobacterium left Earth, sealed in a spore-like state
of suspended animation inside some meteoroid, and
arrived at Mars by the same train of circumstances that
delivered me here. Surely sufficient time has elapsed in
the life of the Solar System for such a sequence of events
to have played out, perhaps even repeated itself.

To read the evidence lining my fractured interior is
to stretch the limits of intuition and instrumentation
alike. High-resolution scanning electron microscope
images show bacteria-like colonies of minuscule sausage-

shaped forms, including one with segments like those of a worm. After close-up photos appeared in news reports worldwide in 1996, however, further investigation suggested that the suspected micro-fossils were neither Martian nor terrestrial remains, but the artefacts of lab procedures that had been used to prepare my samples for study. The processing had caused textural changes that inexplicably copied the outline of familiar life forms – the way a windblown mesa on Mars may perchance assume the contours of a human face.

Three other promising indicators of life, including my content of organic molecules called polycyclic aromatic hydrocarbons, failed to provide conclusive proof. Still remaining to be explained, or explained away, are the tiny grains of magnetite around my carbonate globules. No inanimate process, so far as anyone knows, yields this type of pure magnetite, which is produced on Earth by aquatic bacteria of the strain MV-1. The dark, singular crystals are all that now sustain my proffered hope of Martian life, yet they suffice.

The long-accepted likelihood of Mars as an abode of alternate life draws strength from the planet's solid ground and Earth-like pattern of days and nights. A Martian day – a sol – lasts just a little more than half an hour longer than a day on Earth, since the two bodies spin at nearly the same rate. They also tilt on their axes at almost the same angle, with Mars tipped 25 degrees and Earth 23.5, which accounts for their similar passage through seasons over the course of each year.

The wider compass of the Martian orbit naturally extends each season, as Mars takes 687 Earth days to make its longer, slower way around the Sun. All the seasons are cold ones, with an average annual global temperature of 40 below zero, compared to Earth's 59 degrees Fahrenheit. Prevailing cold, however, need not rule out the possibility of life, considering all the seemingly inhospitable niches on Earth – inside volcanic vents on the ocean floor, in oil reservoirs, in buried rock salt – that serve as homes for tube worms, blue-eyed pink vent fish and other known extremophiles.

The orbits of Earth and Mars bring the two planets within thirty-five million miles of each other every fifteen to seventeen years, enlarging Mars threefold in telescopic views at these times, and thus imposing a natural tempo on the pace of early discovery. With Mars near in August of 1877, for example, its long-suspected moons revealed themselves at last as two small, dark companions, Phobos and Deimos, practically at the limit of detection, and travelling so fast that lunar months on Mars could be reckoned by these satellites in a matter of hours.

During that same 1877 planetary encounter, networks of straight-sided Martian *canali* were sighted from Italy and plotted on new maps of Mars. The Italian 'channels' translated into English 'canals' in time for the next close approach, in 1892, when an American enthusiast insisted he saw several hundred canals, and soon attributed their existence to the desperate irrigation efforts of a dying race.*

The fixed idea of a Martian alter-ego guided preparations for the next Mars approach of August 1924, when civilian and military broadcasters proposed a three-day radio silence, in order to listen for intelligent signals from Martians. The US Army designated its chief signal officer to attempt decoding any intercepted transmissions, and although the value of his services went untested in that effort, British and Canadian wireless operators reported several unidentified radio beeps. Meanwhile, observers in the Swiss Alps directed a greeting to Mars in the form of a lens-amplified light ray reflected off the snow-covered slopes of the Jungfrau, and astronomers confirmed that the moving bright spots seen through improved telescopes were clouds in the Martian atmosphere.

Rather than wait decades for the planets to align themselves in such research-favourable positions, planetologists and rocket scientists of the 1960s began taking advantage of ideal launch opportunities, which arise every twenty-six months, to send a series of flybys, orbiters and landers to Mars. These spacecraft followed the efficient, purposeful route of the Hohmann transfer

* See Percival Lowell, *Mars* (1895), *Mars and its Canals* (1906), and *Mars As the Abode of Life* (1908).

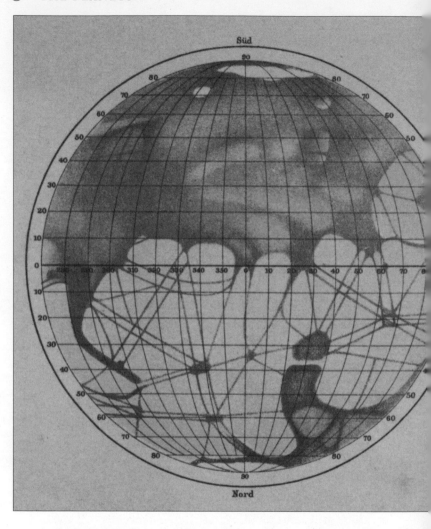

As Schiaparelli studied the Martian surface over time, he noted that most of the planet's canals took on a double form. He suspected

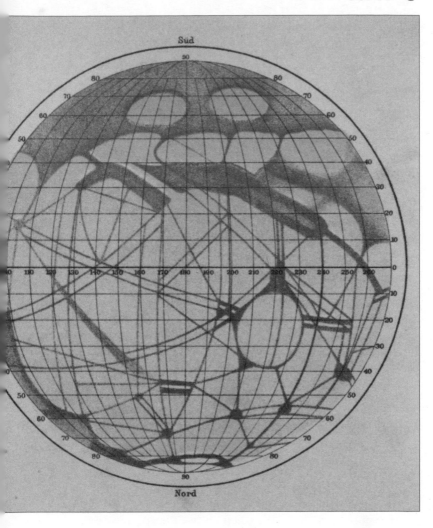

the single straight channels to be parallel double lines as well,
though his telescope could not resolve them.

orbit trajectory, calculated to carry them from Earth launch to intersection with Martian orbit less than a year later, just in time to intercept the planet at that point. *

Mishaps have prevented half the Mars-destined craft from either gaining their difficult objective or functioning profitably there, including three intended landers that accidentally crash-landed and destroyed themselves on contact. Among the numerous successes, however, five landers set up both stationary and roving field laboratories, automated to take in Martian air and soil samples for analysis.

Vikings 1 and *2*, the first pair of twin robot scientists from Earth to search for life on Mars, reached the golden plains of Chryse and Utopia in the summer of 1976, while I lay still entombed in the wintry Antarctic. They settled on landing sites named for classical fantasies and vague nineteenth-century impressions of my native world. Even now that on-site surveys have made sense of Mars's actual topography, many romantic allusions persevere in the otherwise logical nomenclature scheme instituted by modern areographers. Thus the large dry river valleys discovered in the early 1970s, such as Ares Vallis and Ma'adim Vallis, recognize the war god Mars or the word 'star' in various human languages – the sole exception being Valles Marineris, the greatest valley of all, which honours its discoverer, *Mariner 9*, the first

* See S. Glasstone, *The Book of Mars*, NASA Special Publication 179 (1968).

artificial satellite ever to orbit a planet other than Earth. The smaller valleys take their names from Earthly rivers, either classical or actual. (Evros Valles, near my former home, shares its name with a river in Greece.)

Large, ancient Martian craters, newly visualized, now bear the names of scientists and science-fiction writers, including Burroughs and Wells, and small craters the names of small Earthly villages with populations of fewer than 100,000. On the smallest level, individual surface rocks spotted in the intimate photos taken by landed spacecraft have assumed whimsical names from cartoons and storybooks, including Calvin and Hobbes, Pooh Bear and Piglet, Rocky and Bullwinkle, or nicknames based on their appearance: 'Lunchbox', 'Lozenge' and 'Rye Bread'. Although my own name is specific and descriptive, occasionally I have been called 'Big Al' and other such convenient nicknames in closed-door discussions among researchers.

By now some spacecraft have served such lengthy periods of active duty at Mars and relayed such steady streams of information as to enable Earthbound geologists and climatologists to monitor trends over time, notably the transient nature of the Martian polar caps. At the start of every autumn in the south, as much as one third of the atmosphere sifts like powdery snow from the salmon-coloured sky in a white frost of carbon dioxide. The dry ice fluffs the south polar cap a yard thicker and coats the southern hemisphere halfway to the equator all through the winter, the south's longest season. When spring comes, the white rime sublimes directly back into the atmosphere without pausing to melt. Soon it deserts the sky again, precipitating onto the north pole with the arrival of autumn there.

In other studies, Mars-stationed spacecraft have tested the strength of my planet's gravity field, measured the atmospheric content and pressure, clocked wind speeds, compared the heights of mountains to the depths of basins, listened at ground level for Marsquakes, and also detected an iron core, solidified now, and no longer capable of generating a magnetic field.

Indeed, so many spacecraft currently share the Martian domain, returning so many thousands of images, that the picture of the planet grows constantly more refined and more complex to Earthly eyes, with new theories adduced accordingly, so that the controversy among planetary scientists escalates as missions proliferate.

From a Martian perspective, the sum of all this scrutiny could be construed as a hostile invasion.* The Earth envoys, however, have found no entity sensitive to assault, and only the slightest, most equivocal suggestion of any biological activity. The reddish dirt of Mars, rich in iron peroxide and other oxidizing agents, routinely sterilizes itself and all new arrivals. Organic compounds

---

* See H. G. Wells, *The War of the Worlds* (1898).

carried to the Martian surface on meteorites or visiting spacecraft are destroyed at once by the highly reactive chemistry of the present era. Any organic material that survived the chemical attack would no doubt succumb to physical dismantling by the Sun's ultraviolet radiation, since the Martian atmosphere provides no protection comparable to Earth's layer of ozone.

Astrobiologists insist that life on Mars, like the once-plentiful water on Mars, could simply have gone underground to avoid these dangers, and may yet be discovered, extant or extinct, through diligent pursuit. Astronomers agree, asserting that even if Mars ultimately proves void of life, its unique environment will continue to lure robotic and human explorers to its frozen shores.

Some visionaries see in Mars a potential homestead on a high frontier, awaiting colonization.* Scientifically feasible programmes for 'terraforming' Mars to enhance its Earthly likeness propose the fabrication of suitable habitats by, for example, heating the Martian south pole with huge space-based mirrors that would focus and magnify the Sun's light, forcing the residual polar cap of carbon dioxide to sublime like a geyser of greenhouse gas. In the ensuing warmth, pure drinking water might pour from the ice at the north pole, or be mined from

* See Arthur C. Clarke, *The Sands of Mars* (1951), Robert A. Heinlein, *Red Planet* (1949), and Kim Stanley Robinson, *Red Mars* (1993), *Green Mars* (1995), *Blue Mars* (1997).

the abundant buried permafrost, or chemically extracted from select areas of the planet's hardened crust.

Planners say they can achieve the same effect another way, by preparing a safe environment for a few hardy strains of microbes, and releasing them into the Martian regolith, there to ingest available nutrients and excrete gases, including ammonia and methane, which would then thicken the atmosphere, enabling it to hold in more heat, thereby raising the ambient temperature to create a shirtsleeve environment.

Proponents of interplanetary manifest destiny expect that whether or not Mars has ever been inhabited by sentient Martians, Earthlings will eventually become Martians.*

I picture them on the pitiless surface, dressed in specially engineered Mars protection suits, living in domed modules, toiling under an artificially generated magnetic field that shields them from harmful cosmic rays as they harness the energy of the wind and convert local stores of heavy hydrogen to electric power. As they busy themselves in the desert, raising food crops in greenhouses and prospecting for troves of high-grade mineral ores, they continue their careful reconnaissance of the planet, travelling overland by tractor and on foot, scaling and spelunking, still half hoping, half fearing that they are trespassing.

* See Ray Bradbury, *The Martian Chronicles* (1950).

I suppose it is their condition of being alive, and their sense of living such short lives, that drives their obsession to seek other life in every possible redoubt. Even if they succeed in preparing the way for their compatriots to join them in founding a great Martian civilization, they will continue to strain for traces of whatever might have scrabbled in the reddish dust before they arrived.

8

ASTROLOGY

When Galileo, a Pisces with Leo rising, turned his spyglass to the dark over Padua in the winter of 1610, 'guided', he said, 'by I know not what fate', the planet Jupiter appeared to him, bearing four new moons no man had ever seen before.

Galileo thanked God for granting him these sights, and praised his new spyglass as the means. But surely the alignment of the planets through those January nights had also favoured his success. For Venus, along with Mercury, hid below the horizon. Saturn set early in the evening, and by the time Mars rose, three hours before dawn, cold and fatigue had long since forced Galileo indoors. Even the Moon, though almost full at the start of Galileo's vigil, gradually withdrew, leaving bright Jupiter, aglow at opposition, to wander the stars alone.

No sooner had Galileo discerned the planet's four companions than he saw what they augured for his own future: He might gain a position at the Tuscan court by naming them for his most important patron, the young Florentine Prince Cosimo de' Medici. Given Jupiter's prominence in Cosimo's horoscope, which Galileo had already cast, the four moons must represent the boy and his three younger brothers, and therefore should be known henceforth as the Medicean stars.

'It was Jupiter, I say,' Galileo reminded Cosimo, 'who at Your Highness's birth, having already passed through the murky vapours of the horizon, and occupying the midheaven' – by which he meant Jupiter had risen to the dominant, most auspicious position in the sky according to Renaissance astrology – 'and illuminating the eastern angle' – that is, affecting the ascendant sign – 'from his royal house' (Jupiter being considered king of the planets), 'looked down upon Your most fortunate birth from that sublime throne and poured out all his splendour and grandeur into the most pure air, so that with its first breath Your tender little body and Your soul, already decorated by God with noble ornaments, could drink in this universal power and authority.'

Jupiter had thus conferred on Cosimo the expansive confidence and noble ethical concern that befitted a born leader. The positive effect of Jupiter, called 'the greater benefic' by practitioners of the stellar art, was known to uplift a person from pettiness to greatness, as well as to promise health and sanity, levity, wisdom, optimism and generosity.

The two telescopes mounted here, one made of wood, paper and copper, the other of wood covered in gold-tooled leather, are believed to have been built by Galileo. The elaborate ebony frame at the bottom contains the lens that first enabled him to see the moons of Jupiter, and which he later presented as a gift to the Grand Duke of Tuscany.

149

'Indeed,' noted Galileo,

> it appears that the Maker of the Stars himself, by
> clear arguments, admonished me to call these new
> planets by the illustrious name of Your Highness
> before all others. For as these stars, like the offspring
> worthy of Jupiter, never depart from his side except
> for the smallest distance, so who does not know the
> clemency, the gentleness of spirit, the agreeableness
> of manners, the splendour of the royal blood, the
> majesty in actions, and the breadth of authority
> and rule over others, all of which qualities find
> a domicile and exaltation for themselves in Your
> Highness? Who, I say, does not know that all these
> emanate from the most benign star of Jupiter, after
> God the source of all good?

The uproar that followed Galileo's announcement of
his discoveries caused a few commentators to wonder
aloud how the four new celestial bodies would affect
astronomy, on the one hand, and astrology on the
other.

Soon the Medicean stars weighed in as astronomical
evidence to support the unpopular heliocentric system
of Copernicus. By showing they could circle Jupiter even
as Jupiter continued his own heavenly rounds, the new
satellites made plausible the idea of the Earth's moving
through space, together with its Moon, around the Sun.

Astrology broke with astronomy at this point, forced

by its focus on human experience to retain the geocentric outlook. Nor did astrologers see any need to assign a new sphere of influence to the Medicean stars. Rather, they continued to esteem only Earth's Moon, which they regarded as the ancient, familiar, feminine controller of emotional responses and everyday patterns of activity.

In Galileo's own natal chart, for example, the Sun is in Pisces,* but the Moon lies in the sign of Aries at the mid-heaven, indicating a highly imaginative, self-reliant, independent and inventive individual with a restless mind, someone who goes beyond existing boundaries as a pioneer, an adventurer, even a sky warrior. At the same time, the Moon occupies the ninth of the twelve mundane houses – the house ruled by Jupiter and traditionally associated with knowledge and understanding. The Moon in the ninth house signifies strongly held religious and philosophic beliefs, as well as an advanced education and a long-lived mother, all of which Galileo had. The ninth house also encompasses travel to foreign countries, and although Galileo never left Italy, it could be argued that his telescope carried him on the farthest possible journeys.

The same Jupiter that swam as a small globe in the

---

* Two horoscopes drawn for Galileo during his lifetime (1564–1642) show his Sun near six degrees in Pisces. While his birth in Pisa on 15 February would seem to make him an Aquarian (the Sun sign of those born 20 January–18 February), the calendar reforms of 1582 moved his birthday to the 25th.

eyepiece of the spyglass resided, in Galileo's horoscope, in the sign of Cancer – where astrologers say the planet is 'exalted', or most free to express itself through the individual's experience – and also conjunct with Saturn in the twelfth house. Jupiter and Saturn aligned in the house of confinement spelled success for Galileo around the age of forty or fifty. (He was forty-seven when he published the astronomical findings that brought him instant fame.) Together, Jupiter and Saturn implied that Galileo would face ideological crises (such as his later clash with the Inquisition, perhaps) and live in seclusion and solitude (as he did under house arrest his last eight years). The ebullient increase and fertility of Jupiter is tempered, in Galileo's nativity, by the sobering nearness of Saturn.

Jupiter assumed its astrological mantle of benevolence and largesse in Babylonian times, around 1000 BC – long before Sir Isaac Newton (a Capricorn) grasped the planet's true physical enormousness by watching it pull on Galileo's moons. The ancients had no way to assess the sizes of the planets or the distances between them, so their association of Jupiter with grandeur poses a mystery for astronomy and astrology to share.

As befits the planet of expansion, Jupiter more than doubles the mass of the other eight planets combined. Compared to the Earth alone, Jupiter measures 318 times Earth's mass, and 1,000 times Earth's volume. The diameter of Jupiter, however, is 'only' eleven times that of Earth, since the giant compacted itself as it accreted, so its diameter expanded at a fraction of the rates at which its mass and volume increased.

A world apart from the terrestrial planets, Jupiter mimics the Sun in both composition and attitude: it consists almost exclusively of hydrogen and helium, and reigns over its own replica solar system of at least sixty planet-like satellites – the four largest ones Galileo found, plus fifty-nine others discovered (so far) since the dawning of the Age of Aquarius.

Although many of Jupiter's moons are rocky bodies, the gas giant itself has no solid surface, no terrain of any kind. The face it presents to Earthly observers is an expanse of pure weather: every identifiable feature resolves into a cloud bank, a cyclone, a jet stream, a thunderbolt, or a curtain of auroral lights. On Jupiter, a storm may continue for centuries, innocent of landfall. No seasonal changes disrupt the weather patterns either, since the planet stands erect on its axis, only three degrees atilt.

Countervailing winds that tear east and west across Jupiter arrange its clouds in a canopy of horizontal stripes. The east-flowing jet streams alternate with the westward trade winds to form some dozen dark belts and bright zones, each one confined to its own latitude band, where it remains fixed over time. Generations of Jupiter watchers have marvelled at the persistence of these neat divisions.

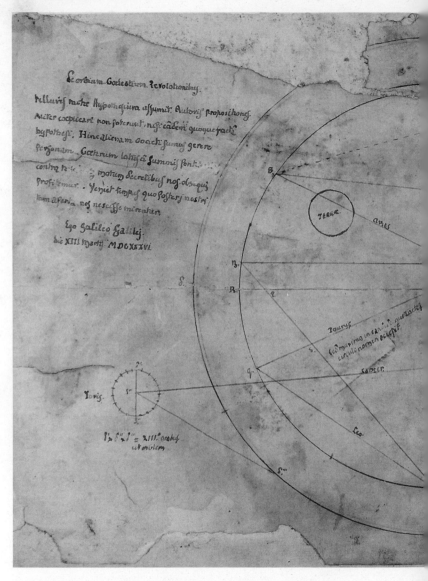

An adept astrologer, Galileo noted planetary positions to cast horoscopes for himself and his children, as well as for his royal patrons.

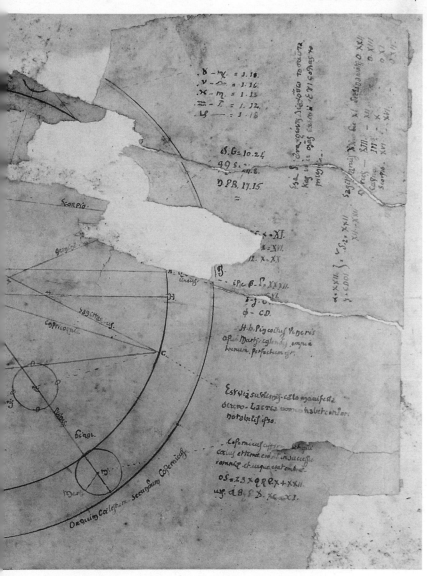

Every band of wind hosts a meteorological drama within its bounds. In the South Equatorial Belt, for example, a stable, oval-shaped storm known as 'the Great Red Spot' has been studied continuously since 1879. The Spot has faded from its once vivid vermilion to pale orange, and shrunk to half its former width (though it still exceeds the diameter of Earth) without ever changing lanes. When the Great Red Spot meets other clouds travelling faster or slower in the same direction in the same belt, it sweeps them up and keeps them circling its perimeter for weeks, until they either merge with it or whirl past. Small oval storms that form in the dangerous furrows between east- and west-rushing flows, however, quickly fall victim to shear forces and shred apart in a day or two, like downgraded hurricanes.

Jupiter's clouds take their red, white, brown and blue colours from sulphur, phosphorus and other impurities in the atmosphere. Winds marble the cloud colours, as though with an eye to beauty, and eddies feather the edges of designs. All the colours might well have blended and muddied by now, after aeons of swirling, had not each pigment generally held fast to its own layer at a designated altitude in the atmosphere. The low-down cloud base of warm blues can be glimpsed only through breaks in the overlying browns and whites, which give way, a few hundred miles up, to the high-flying cold reds.

A faint but detectable glow of infrared radiation leaks

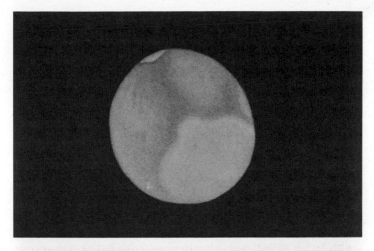

Observing the planets from his home at Slough, Sir John Herschel drew Jupiter as it appeared through his 20-foot reflector on 23 September 1832. He noted that astronomers were able to time the planet's rotation rate at 9 hours 55 minutes and 21 seconds by paying careful attention to the features in the belts.

through gaps in the cloud cover. This is the lingering heat of the planet's original accretion, rising slowly by convection from the core as Jupiter continues to cool and contract. Half a billion miles from the Sun, Jupiter releases more warmth than it receives. Most of the energy to drive the Jovian winds thus derives from within, augmented only slightly by faint Sunlight falling from afar. Jupiter's radiance has earned it a reputation as a 'failed star', but its internal temperature, estimated at 17,000 degrees, falls very far short indeed of the fifteen-million-degree inferno that makes the Sun shine.

The vast, variegated clouds, which are all anyone ever sees of Jupiter, constitute only a thin veneer surrounding the planet; they comprise less than 1 per cent of its forty-five-thousand-mile radius. Underneath the clouds the atmosphere grows denser and hotter because of mounting pressure, and the weather stranger. Here the carbon content of methane and other trapped gases may be crushed to tiny diamonds in the sky. Gradually the gases cease to behave as gas, as they dissolve into a sea of liquid hydrogen.

Some five thousand miles down into this milieu, where the pressure reaches at least a million times Earth's norm, the liquid hydrogen turns opaque, metallic, molten and electric. By far the greatest part of Jupiter consists of hydrogen compressed to this exotic phase.

According to astrological lore, each planet corresponds to a specific metal, so that silver, for example, pairs with the Moon, gold with the Sun, and mercury with Mercury. Jupiter's assigned metal has been tin, not hydrogen. But then, no medieval alchemist knew of hydrogen's existence, let alone the bizarre concoction of liquid metallic hydrogen produced inside Jupiter.

Modern scientists have fabricated only the minutest quantities of liquid metallic hydrogen, by means of reverberating shock waves inside laboratory apparatus, and each such painstakingly made sample lasts just one-millionth of a second. Nevertheless, theorists have gleaned the essence of the substance and, by extrapolation, explained many aspects of Jupiter's nature. Its magnetic field, for example, which is twenty thousand times the strength of Earth's field, and extends all the way to the orbit of Saturn, arises from the liquid metallic hydrogen interior. A genuine Jovian dynamo is created deep inside the planet, where warm currents of escaping heat stir a susceptible fluid shot through with electric currents generated by Jupiter's rapid rotation.

The whole mammoth bulk of Jupiter rotates in just under ten hours, faster than any other planet. Its massive body honours the memory of the Solar System's earliest beginnings as a spinning disk, and none of Jupiter's attendant moons can slow it down. As to the giant's rate of revolution in orbit, however, its far remove from the Sun relaxes its pace and adds many miles to its annual travels.

At five times the Earth-Sun distance, Jupiter takes a

long year, the equivalent of twelve Earth-years (eleven years and 315 days), to orbit the Sun. En route it spends about one Earth-year passing through each of the twelve zodiac constellations. In traditional Chinese astrology, Jupiter's slow gait earned it the title of 'Year Star' (*Sui xing*) – the determiner of the Chinese years of the rat, ox, tiger, rabbit, dragon, snake, horse, sheep, monkey, rooster, dog and pig. The Chinese cycle of animals, however, bears only scant relation to the twelve signs of the western zodiac, which include a bull, a lion and a crab, as well as half-human twins, a virgin and a water-bearer.

In western astrology, one or another planet 'rules' the sign with which it shares a natural affinity. Jupiter, long regarded as the most fortunate planet, rules Sagittarius, the archer, the sign of people born from mid-November to mid-December, who are said to express themselves with open-minded vision and honesty. For many centuries Jupiter also ruled the sign of Pisces, the fish, whose February–March natives (including Galileo) are masters of memory and introspection. But then, after the discovery and naming of Neptune in 1846, the new planet became astrologically associated with water, and so took Pisces away from Jupiter.

Unlike dim, distant Neptune, Jupiter makes such a naked-eye spectacle of golden light in the night sky that its presence has been known since antiquity, and therefore its discovery cannot be dated. Although the time of Jupiter's birth has been deduced, the place of birth may lie far beyond the region where the planet now resides.

Planetary astronomers say that Jupiter formed 4.5 billion years ago from a seed of rock at a fortuitous location that predisposed it to gigantism. Far from the proto-Sun, the proto-planet rolled through the cold reaches of the primordial nebula, gathering icy tufts of hydrogen-rich compounds such as methane, ammonia and water. Upon quickly attaining ten or twenty times Earth's mass, young Jupiter drew in the nebula's still plentiful light gas and grew fat on hydrogen and helium.

No small world could have retained such a large envelope of gas, but Jupiter succeeded because of its superior mass and consequently stronger gravity. Jupiter's attractive power, the strongest of all the planets, also detoured passing comets from their elongated paths around the Sun and forced them into Jovian orbit instead. It was most likely by consuming some number of these comets that Jupiter augmented its stores of carbon, nitrogen and sulphur.

The whole world witnessed one such comet capture when Periodic Comet Shoemaker-Levy 9 crashed through the Jovian cloud banks. In 1992, this comet brushed so close to Jupiter that the planet tore it into twenty-one chunks as big as icebergs, plus many more as small as snowballs. The pieces then circled Jupiter for two years in single file, like a flying string of pearls, before diving to their destruction, one by one, over a week's time in mid-July of 1994. As they fell through Jupiter's atmosphere, they exploded in fireballs and thousand-mile-high plumes of debris.

Each detonation left a huge bruise on the clouds, until a whole necklace of black pearls hung around Jupiter, just south of the Great Red Spot. Although every comet fragment had struck the planet's far side, out of telescope range, rapid rotation soon carried each new impact into view. The dark stains then spread thin on shock waves and winds, dispersed from day to day, and all but disappeared by late August, before scientists could distinguish between the cometary material and the inventory of elements dredged up from the planet.

Following the comet's natural, unintentional probing of the Jovian atmosphere, the *Galileo* spacecraft reached Jupiter seventeen months later, in December 1995, and dropped a robot probe carrying seven scientific instruments through the clouds.

In the single hour it operated before being wrecked by heat and pressure, the *Galileo* probe radioed back eyewitness reports. It found that the high winds seen at high altitudes blew far more forcefully lower down, reinforcing the idea that the winds draw their energy from deep within the planet. The probe also measured relatively large quantities of the noble gases argon, krypton and xenon on Jupiter. The abundance of these substances was what forced astronomers to consider a Jovian birthplace far from the planet's present home, out where frozen caches of noble gases could be incorporated into the infant planet. Later on, they reasoned, Jupiter drifted closer in as a result of countless gravitational interactions with other Solar System bodies.

The uniqueness of the on-site vantage point empowered the *Galileo* probe to overturn long-accepted theories by its every discovery. Likewise the things it failed to find caused consternation and conjecture throughout the planetary science community, as when water turned up missing from the returned data.

Astronomers had predicted that the probe, after piercing the visible, colourful ammonia cloud level, would fall through a thick lower layer of ice- and water-laden clouds, where it could be rained on, even struck by lightning. Classical astrologers had also characterized Jupiter as 'moist', in a medieval medical system that claimed the various planets' hot, cold, moist and dry qualities influenced human health by shifting the balance among the four bodily humours – blood, phlegm and black and yellow bile. Moist Jupiter, holding sway over the blood, also inspired the 'sanguine' temperament in individuals, making Jupiterians generally cheerful, or 'jovial', as opposed to mercurial, martial, or saturnine.

Contrary to all expectations, the *Galileo* probe had by chance encountered a dry area, entering a rare hot spot – one of those breaks in the clouds where Jupiter's heat escapes into space. In time, however, the *Galileo* orbiter,

mother ship to the probe, photographed titanic lightning bolts a thousand times brighter than Earthly dis-
charges, and confirmed the presence of atmospheric
water vapour. Indeed, outside the hot-spot 'deserts' that
continually shift their locations around Jupiter, many
parts of the atmosphere appear saturated with water.

The orbiter portion of the *Galileo* spacecraft went on
to explore the Jovian system for seven years. Unlike the
probe, which made only a quick diagnostic descent into
Jupiter, the orbiter became a long-lived artificial companion to the Galilean satellites.

Galileo took commands from mission controllers at
the Jet Propulsion Laboratory in southern California,
who periodically fired the spacecraft's rocket engine to
adjust its orbit, sending it now close in toward Jupiter
to visit Europa, now out on a wide loop to fly by distant
Callisto. As *Galileo* navigated among the Galilean
moons, it discerned the defining characteristic of each:
nearby Io, the reddest, most volcanic body known;
Europa, host to an ice-capped salt water ocean; Ganymede, the Solar System's largest satellite; Callisto, one
of the most primitive and pummelled ones.*

Just as the planetary alignments in a horoscope limn

---

* Johannes Kepler (1571–1630), court astronomer and astrologer
in Prague, first referred to the 'Medicean stars' as 'Galilean satellites'
in 1610. Simon Marius, a contemporary of Galileo and Kepler, gave
the moons their enduring individual names by selecting four
favoured lovers of the mythological Zeus/Jupiter.

the possibilities of a life, so the relative positions of these moons have determined their destiny. Io, the nearest, exhibits the trauma of a too-close attachment. Jupiter's gravitational pull has racked Io with tidal stress, keeping its interior permanently melted, so that fire fountains of lava spew unceasingly from some one hundred and fifty active volcanoes.

Europa, the next nearest to Jupiter and the smallest of the Galilean satellites, also shows signs of internal heating by tidal stress. But the material melted on Europa has apparently been ice, not rock. Thanks to *Galileo*, many scientists now believe a salty sea, more voluminous than the Atlantic and Pacific together, lies sandwiched between Europa's frozen surface and its rocky depths, and moreover that the waters might support some form of extraterrestrial life.

Ganymede, though larger than the planet Mercury and further from Jupiter than Io or Europa, also endures tidal stress. Internal heat keeps Ganymede's iron core partially molten, and this conductive, convective interior sustains the moon's own magnetic field, similar to Jupiter's field, albeit much smaller and weaker.

Only Callisto, beaten and scarred by large ancient impacts, stands aloof from tidal effects. Callisto lies so far from Jupiter that it requires more than two weeks to orbit the planet, while Io makes its way around in less than two days, Europa in three, and Ganymede in seven. Meanwhile the mammoth invisible bubble of the Jovian magnetosphere, which extends millions of miles into space and engulfs all the planet's many moons, spins in synch with Jupiter every ten hours.

As the magnetosphere races past the moons, it bombards them with charged particles, and makes off with fresh particles lifted from their surfaces. The volcanoes of Io pour a constant stream of ions and electrons into the magnetosphere, inducing tremendous currents between Io and Jupiter, several million amperes strong. Indeed, the orbit of Io seethes with so much electrical activity and lethal radiation that it poses a threat even to unmanned spacecraft. *Galileo* had to wait until quite late in its study of the Jovian satellites to risk any close flybys of Io. And every time *Galileo* did pass near Io, one or another of its instruments would shut down, or act up, or take a particle hit that at least partially disabled it. In the end, however, *Galileo* proved so resilient that it once flew *through* the plume of an erupting volcano and survived to recount its experience.

This valiant spacecraft, beset from the outset by numerous difficulties that delayed its launch and threatened its performance, developed a distinct personality that endeared it to the engineers who built it and the astronomers it served. Sometime between 1982 (the intended launch date) and 1989 (the year of the actual launch), *Galileo* suffered damage that went undetected until the craft was well en route to Jupiter. First its umbrella-like main antenna, designed to beam hundreds of thousands of digital images and instrument readings back to Earth, refused to open all the way; then the spacecraft's tape recorder, meant to store data between broadcasts, jammed. Desperate mission controllers worked from the ground for four years to repair and reprogram the star-crossed *Galileo* in space, before it got to Jupiter in 1995. Their efforts not only salvaged the spacecraft, but also prolonged its expected life in orbit, so that the mission was deemed a triumph, even though the communications setbacks reduced the anticipated flood of information to a trickle.

Had astronomy and astrology not parted ways so long ago, some of the *Galileo* mission's problems might have been foreseen. A natal chart drawn for *Galileo*, 'born' at Cape Canaveral on the day of its launch, 18 October 1989, reflects a strong, even aggressive spacecraft, with the Sun in Libra for balance, and Mars conjunct with the Sun at the midheaven, adding ambition. At the ascendant, Saturn, Uranus and Neptune cluster together, which lends a sense of seriousness and importance to the venture. Mercury, however, the planet of communication, makes the worst possible angle – a square, or negative aspect – with Jupiter's position. Another unfortunate Mercury square opposes the powerful triad of Saturn, Uranus and Neptune.

The chart shows Jupiter occupying *Galileo's* seventh house, the mansion of marriage and partnership. Surely the spacecraft partnered with Jupiter through its life work, and also united with Jupiter in its ultimate fate. As the ageing *Galileo* ran out of rocket fuel for steering control, it obeyed one last command that directed it on a collision course for the giant planet. If *Galileo*, with its onboard store of plutonium, were left a derelict in orbit, NASA officials feared, it might one day stray into Europa, contaminating the pristine seas there, or even killing some nascent life form.

On 21 September 2003, the day of its demise, *Galileo* descended into Jupiter's clouds, disintegrated and scattered its atoms to the Jovian winds. 'It has rejoined the probe,' some project scientists said, as though mourning a friend laid to rest. 'They are both part of Jupiter now.'

By the final hour of *Galileo*'s odyssey, the spacecraft's horoscope showed Saturn, the planet of endings, well inside the eighth house, the mansion of death.

9

MUSIC OF THE SPHERES

Between 1914 and 1916, the English composer Gustav Holst created the only known example of a symphonic tribute to the Solar System, his Opus 32, *The Planets, Suite for Orchestra*. Neither Haydn's 'Mercury' (Symphony No. 43 in E flat major) nor Mozart's 'Jupiter' (No. 41 in C, K. 551) had attempted as much. In fact, the title 'Jupiter' did not attach itself to Mozart's work until decades after his death. Similarly, Beethoven's 'Moonlight Sonata' was known for thirty years as Opus 27 No. 2 before a poet likened its melody to moonlight shining on a lake.

The Planets suite contains seven movements, as opposed to nine. Pluto had not yet been discovered at the time Holst was writing, and he excluded Earth. Nevertheless the piece persists as musical accompaniment to the Space Age, partly because people still like it, and partly because nothing else has supplanted it. To make up for its lacks, contemporary composers have augmented it with occasional new movements, such as 'Pluto', 'The Sun', and 'Planet X'.

Holst grew interested in planets through astrology. In 1913, after a burst of reading on the subject, he began casting friends' horoscopes and thinking of the planets in terms of their astrological significance, such as 'Jupiter, The Bringer of Jollity', 'Uranus, The Magician', and 'Neptune, The Mystic'. His daughter and biographer, Imogen, also a composer, recalled that her father's 'pet vice' of astrology led him on to study astronomy, 'and the excitement of it would send up his temperature whenever he tried to understand too much at once. He was perpetually chasing the idea of the Space–Time continuum.'

A natural affinity between music and astronomy has prevailed since at least the sixth century BC, when the Greek mathematician Pythagoras perceived 'geometry in the humming of the strings' and 'music in the spacing of the spheres'. Pythagoras believed the cosmic order obeyed the same mathematical rules and proportions as the tones on a musical scale. Plato reprised the idea two centuries later, in *The Republic*, introducing the memorable phrase 'music of the spheres' to describe the melodious perfection of the heavens. Plato spoke also of 'celestial harmony' and 'the most magnificent choir' – terms that imply the songs of angels, though they referred specifically to the unheard polyphony of the planets in their gyrations.

Copernicus cited the 'ballet of the planets' when he choreographed his heliocentric universe, and Kepler built on the work of Copernicus by returning repeatedly to the major and minor scales. In 1599 Kepler derived a C major chord by equating the relative velocities of the planets with the intervals playable on a stringed instrument. Saturn, the furthest and slowest planet, issued the lowest of the six notes in this chord, Mercury the highest.

As Kepler developed his three laws of planetary motion, he expanded the planets' voices from single notes to short melodies, in which individual tones represented different speeds at given points along the various orbits. 'With this symphony of voices,' he said, 'man can play through the eternity of time in less than an hour and can taste in small measure the delight of the Supreme Artist by calling forth that very sweet pleasure of the music that imitates God.'

For his 1619 book, *Harmonice Mundi* (The Harmony of the World), Kepler drew the five-line musical stave with key-signatures for the several parts, and set down each planet's theme in the hollow, lozenge-shaped tablature of his time. Mercury's highly eccentric, high-speed, high-pitched refrain ranged seven octaves above Saturn's bass-clef rumbling from low G to low B and back again.

'I feel carried away and possessed by an unutterable rapture over the divine spectacle of the heavenly harmony,'

173

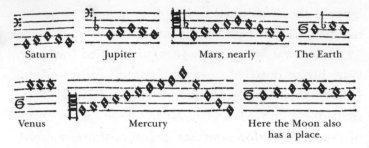

Saturn   Jupiter   Mars, nearly   The Earth

Venus   Mercury   Here the Moon also
has a place.

From the slowest celestial speed of Saturn at aphelion to the
rush of Mercury at perihelion, Johannes Kepler (1571–1630)
interpreted the planets' motions as song.

said Kepler. 'Give air to the heaven, and truly and really
there will be music.'*

The two *Voyager* spacecraft, launched in 1977 and
currently heading for the outer boundaries of the Solar
System, further this musical heritage. As potential
envoys to extraterrestrials, both craft carry a specially
engineered golden record (complete with its own play-
back equipment) that expresses the music of the spheres
as computer-generated tones designating the velocities
of the Sun's planets. The *Voyager* Interstellar Record also
says 'Hello' in 55 languages and plays music selected
from numerous cultures and composers, including
Bach, Beethoven, Mozart, Stravinsky, Louis Armstrong
and Chuck Berry.

Whether by intention or inspiration, Gustav Holst

* Paul Hindemith's 1956–7 opera, *Die Harmonie der Welt* (The Harmony of the World), dramatizes Kepler's work on the planetary
order.

174

ignored the established order of the planets when he initiated his suite with 'Mars, The Bringer of War' in July 1914. Real war, what Holst's generation called The Great War, broke out that autumn, but the forty-year-old Holst, barred from active service by neuritis and near-sightedness, moved directly on to 'Venus, The Bringer of Peace'. In performance, as in composition, the full suite invariably begins at Mars, travels inwards to Venus and 'Mercury, The Winged Messenger', then out again to Jupiter and straight on through Saturn and Uranus to Neptune, where the voices of a female choir, seques-tered in a room offstage, are made to fade out at the finale (with no sacrifice in pitch) by the slow, silent closing of a door.

The suite's immediate popular success amazed Holst, and changed him from an accomplished musician to a famous one. Forced to comment publicly on *The Planets*, he let it be known that 'Saturn, The Bringer of Old Age' – at nine minutes forty seconds, the longest of the suite's seven movements – was his favourite. 'Saturn brings not only physical decay,' Holst said in the planet's defence, 'but also a vision of fulfilment.'

Seen for the first time through a backyard telescope, ringed Saturn, icon of the other-worldly, is the vision most likely to turn an unsuspecting viewer into an astronomer for ever. The spectacular Saturnian ring system spans a disk 180,000 miles wide from one ring tip, or ansa, to the other. Its vast breadth approaches the distance from Earth to the Moon, yet the average ring depth scarcely exceeds the height of a thirty-storey building. In Holst's day, astronomers trying to describe the rings' incomparable flatness grasped at pancakes and gramophone records as metaphors, before settling on a sheet of thin cardboard the size of a football stadium. (Improved measurements have since replaced the cardboard with tissue paper.)

Saturn appears with Jupiter and Venus in a painting of the night sky over Holst's beloved Cotswolds, given to him at the 1927 festival in his honour where he conducted *The Planets* for the last time. The artist Harold Cox said he had consulted the Astronomer Royal on the correct placement of the planets for this portrait of a May night in 1919 – the year the public first heard *The Planets* in concert, and Holst won appointment as professor at the Royal College of Music. Saturn looks like a mere bright spot in the painting, duller than the lights of Jupiter or Venus, and ringless, of course, since the naked eye cannot discern the celebrated rings. This is not to say they are invisible or absent from the painting, however. On the contrary, the rings sparkle so with reflective ice and snow that they fairly triple Saturn's lustre. All the ring components, which range in size from dust grains to boulders big as houses, are thought to be at least ice-coated, if not wholly composed of frozen water. The body of Saturn, in contrast, is a gas giant much like Jupiter, made of hydrogen and helium, though smaller and paler and twice as far removed from the Sun. Without its surround of ice crystals, snowflakes, and snowballs of all sizes, Saturn would hardly dazzle viewers a billion miles away.

In May 1919, the rings were tipped towards Earth to Saturn's artistic advantage. Approximately once every fifteen years, or twice during Saturn's 29.5-year orbit of the Sun, the rings turn edge-wise to earthly admirers, and withdraw their flattering light. Even in telescope views, all that can be seen of the rings at such times is a thin shadow line across the planet's yellowish globe. Such periodic disappearances confounded the rings' earliest observers.

Galileo, the first to glimpse bulges alongside Saturn in July 1610, mistook them for a pair of close 'companions', which did not move about like Jupiter's satellites, but hugged the planet's flanks to make it appear 'triple-bodied'. Monitoring Saturn over the next two years, Galileo confessed astonishment in late autumn of 1612 to find the planet suddenly solitary and round, deserted by its erstwhile supporters. 'Now what is to be said about such a strange metamorphosis?' he wrote to a fellow philosopher. Perhaps the planet Saturn, like its mythological counterpart, had 'devoured his own children'?

Galileo predicted the companions would return, and when they did they were much altered. In 1616 he said they resembled a pair of handles on Saturn, and later he likened them to ears, though he never grasped the

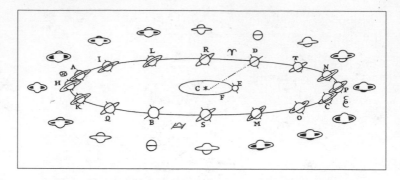

Christiaan Huygens (1629–1695) prepared this diagram for the 1659 publication of his book, *Systema Saturnium*, to show how Saturn's appearance changes, over the course of its nearly thirty-year orbit, in the eyes of Earthly admirers.

fantastic nature of their true identity. Not until 1656 did the Dutch astronomer Christiaan Huygens lay the changing shape of Saturn to the existence of 'A broad, flat ring, nowhere touching, and inclined to the ecliptic.' Huygens published a full explication in his book, *Systema Saturnium*, in 1659.*

Huygens always spoke of 'Saturn's ring' as a single solid entity, and so it was thought to be until 1675, when Jean Dominique Cassini, director of the Paris Observatory, detected a dark dividing line that split the ring into two concentric lanes, dubbed 'A' (for the outer one) and 'B' (the inner and brighter). The passage of

* Both Galileo and Huygens were able lute performers and friendly with many composers. Huygens also experimented with a 31-tone equal-temperament scale that influenced the music of the Netherlands into the twentieth century.

another two centuries yielded a third segment – the dim interior 'C' ring, discovered in 1850 – though still no one could ascertain how any of the rings was made. Embattled opinion on ring structure swayed from solid sheets to swarms of small satellites to rivers of orbiting liquid and exhalations of planetary vapours.

'I have effected several breaches in the solid ring,' young James Clerk Maxwell of Scotland boasted in 1857 from the midst of his mathematical calculations, 'and am now splash into the fluid one, amid a clash of symbols truly astounding.' Convinced that Saturn's gravity would shatter a solid construction of such large dimensions, Maxwell construed the rings as a profusion of individual particles so numerous as to create the far illusion of solidity. Each particle would perforce pursue its own orbit, according to Kepler's laws, with the particles furthest from Saturn travelling at the slowest speeds, and the nearer ones more quickly, just as Saturn itself proceeded ponderously about the Sun compared to the rapid pace of Mercury. (What choruses Kepler might have scored for these multitudes!)

Within the crowded rings, particles constantly jostle their neighbours, bumping each other into wider or narrower orbits by the exchange of energy and momentum. Collisions also fling particles above and below the flat plane of the rings, but such strays are quickly whipped back into line.

Since 1966, four more rings, designated D through

179

G, have joined Saturn's classical A, B and C rings. As a group, they flout the alphabetical order of their discovery in their progression outwards from Saturn – D, C, B, A, F, G, E – like notes on a practice scale. Each lettered region distinguishes itself from the others by slight colour or brightness variations, or the density of its particles, or its unusual shape. When seen from the privileged perspective of a visiting spacecraft, the lettered segments further resolve into myriad slender ringlets, separated by as many tiny gaplets, and patrolled by embedded moonlets.

The ring system probably formed from the break-up of an icy moon, or perhaps a captured planetoid, some sixty miles in diameter. That hapless body, destroyed a few hundred million years ago, may still be striving to reassemble itself in Saturn's orbit. As its particles gravitationally attract one another and cling together, they form larger aggregates that pull in additional particles to keep on growing bigger, but only up to a point. Any accreting ring body that exceeds certain size limits gets torn apart by Saturn's tidal forces, and so the scattered bits seem destined never to reunite into a single satellite.

Earth's Moon, which passed through a similar stage as a ring of collision debris, nevertheless cobbled itself together because its pieces orbited far enough from our planet to escape the destructive effects of tidal forces. At Saturn, the rings huddle close. They occupy a nearby region of perpetual fragmentation known as the Roche zone, named for the nineteenth-century French astronomer Edouard Roche, who formulated the safe distances for planetary satellites. The larger moons of Saturn all lie well beyond Roche's limit, outside the perimeter of the rings. However, Saturn's extended family (at least thirty-four moons at last count) includes many small members in and among the rings that help sculpt their intricacies. The F ring, for example, owes its peculiarly twisted and narrow outlines to the action of two accompanying moonlets, one of which runs rapidly along the inside of the ring while the other laps its outside. Together they work as 'shepherd' satellites, herding the flocks of particles between them into clumps, knots, braids and kinks.

When the *Cassini* spacecraft reached Saturn in the summer of 2004, it trumpeted its arrival by soaring up through the gap between the F and G rings, skimming across the broad expanse of the ring plane, and then diving back down through the far side of the same gap, where it emerged unscathed. The relative emptiness of such spaces results from the interplay of Saturn's satellites with particles in the rings, following the same rules Pythagoras defined in his experiments with strings.

Pythagoras had shown how the pitch of a string rose an octave when he shortened its length by half. Playing strings of these two lengths together pleased the ear, he said, because their vibrations resonated in the whole-number relationship of 2:1. Other whole-number relationships, or resonances, yielded other felicitous musical intervals, such as thirds, fourths and fifths. Galileo, commenting on the effects of sympathetic vibrations in his book *Two New Sciences*, judged that the octave 'is rather too bland and lacks fire', while the sound of a 3:2 resonance (the musical interval of the fifth) caused 'a tickling of the eardrum so that its gentleness is modified by sprightliness, giving the impression simultaneously of a gentle kiss and of a bite'.

The most notable resonance effect in the rings of Saturn is the Cassini Division – the three-thousand-mile-wide separation between the A and B rings. The Division derives from its 2:1 resonance with the moon Mimas, orbiting more than 40,000 miles away. Ring particles within the Cassini Division travel twice around Saturn to Mimas's once, and so they repeatedly overtake the slower-moving moon at precisely the same two points in their orbit. There they gravitate towards it. Eventually the pull of the moon, boosted by rhythmic repetition, boots the particles out of the resonant orbit, clearing the gap. A similar but narrower gap near the outer margin of the A ring, called the Encke Division (for Johann Encke, a former director of the Berlin Observatory), shares a 5:3 resonance with Mimas and a 6:5 resonance with another moon. Also, the decorative scalloped border on the outer edge of the A ring owes its six petal-like lobes to a 7:6 resonance with two small satellites that occupy a single orbit and may once have been a single object.

The rings resonate also to the beat of Saturn's rapidly rotating magnetic field. Generated within the planet's liquid-metallic-hydrogen interior, the magnetic field spins in time with Saturn's rotation every 10.2 hours. Particles in the B ring travelling just that fast – or half as fast, or twice as fast – are consequently driven from their orbits.

Saturn reigned as the lone ring world for three hundred years, until the discoveries of the 1970s and 1980s showed that all the giant planets bear rings of some kind. Jupiter has tenuous, transparent 'gossamer' rings consisting of dross flaked off the surfaces of several small moons. Uranus owns nine dark, narrow rings with sharply defined borders constrained by shepherd satellites. And Neptune's five faint, dusty rings are so irregular in thickness that some sections thin almost to nothingness, leaving the impression of partial ring arcs. None of these recently recognized ring systems can really compete with Saturn's baroque, even rococo rings. Rather, each of the others portrays a single nuance of ring dynamics – some phenomenon present at Saturn as well, but drowned there in the volume of variations and embellishments.

All the rings undergo constant change through repetitive rounds of build-up and breakdown. From year to year, they are the same yet not the same. As they fray and wear away in the friction of internal collisions, new infusions of moon dust and infalling meteorites replenish the particle supply.

Each ring system, the product of gravity and harmony, suggests a template for cosmic design. Rings recall the birth of our whole family of planets, which arose from the flat, spinning disk that surrounded the infant Sun five billion years ago. Rings also find frequent echo now in the so-called 'protoplanetary disks' discerned around distant young stars, where the raw materials of gas and dust are uniting in new world syntheses. Saturn's rings thus link our Solar System to other, extrasolar systems in the making, and the present Solar System to its own ancient past.

'Music,' Holst observed in a letter to a friend, 'being identical with heaven, isn't a thing of momentary thrills, or even hourly ones. It's a condition of eternity.'

10

DISCOVERY
OR
'NIGHT AIR'

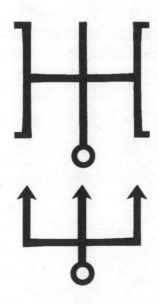

The Herschels worked a great many years. Sir William Herschel's papers, published in various scientific journals, stretch through a period of forty years. Sir John Herschel's reach through a period of fifty-seven years – about twice the average length of life. Sir William Herschel died at eighty-three, Sir John at seventy-eight; and, as if to show that a woman can live and work even longer than a man, Caroline, the sister of Sir William, died at ninety-eight.

Is it worth while to talk about the unhealthiness of 'night air' when that class of people who are most exposed to its influence, whose calling keeps them breathing it, are so long-lived? For the work of the practical astronomer is mainly out-of-doors and in good night air, instead of indoors in bad air. (I think it is Florence Nightingale who asks, What air can any one breathe in the night except night air?)

Maria Mitchell, American astronomer (1818–89)

Hanover, Germany, November __, 1847

My Dear Miss Mitchell,
Please accept my most excited congratulations on your recent discovery. Word of 'Miss Mitchell's Comet' had already reached me from several sources here on the continent before your letter arrived, as well as from my nephew* in London, but how delighted I am to know that you thought of me in your hour of glory, and took the time to share your triumph with an old woman. Indeed, as you say, you and I enjoy a special bond. Even though my own telescope is now the chief ornament of my sitting room, it let me watch many comets come out of the dark, looking dull and plainly clad at first, but growing as they approached, until, nearing the Sun, they sprouted their great fuzzy caps and spread the tails that are the peacocks of the cosmos.

I am particularly gratified to hear the new comet will keep your name, Miss Mitchell, because such fame will secure your future as nothing else could do. One of my comets took Professor Encke's name, after he computed its orbit and predicted its return.† That still left me seven other 'lady's comets', though I had no need of any, what

---

* Caroline Herschel's nephew, Sir John Herschel (1792–1871), was president of the Royal Astronomical Society and son of renowned astronomer Sir William Herschel (1738–1822), who discovered the planet Uranus in 1781.
† The comet named for Johann Franz Encke (1791–1865), who became director of the Berlin Observatory in 1825, returns every 3.3 years.

with my brother's name about me like an aegis, plus a royal pension as his assistant. You, however, are a young woman alone in a young country, and the discovery of your comet surely surpasses your employment at the Nantucket Library, both in terms of assuaging your family's concerns for your welfare AND rousing the world's regard for your abilities.

Just the way your father, bless him, has encouraged your pursuits, so did my brother support me in mine, although I suppose the more correct assessment is that he trained me because he required an adept assistant willing to struggle long hours at his side as no hired, indentured, or enslaved help would do. The irony is that while I became William's right arm in his astronomical investigations, and kept all the official nightly records, I was ABSENT that particular evening during the week of my birthday when he discovered the 'comet' we are now pleased to call the planet Uranus.*

William was not seeking a planet, of course, for we held it almost an article of faith that only six planets orbited the Sun. When his sweeps of the heavens turned up something blurred or indistinct, something that stood out from the stars' points of light, he naturally wondered whether he had come upon a new comet he could claim as his own, or someone else's comet on a

---

* Sir William first noted what proved to be Uranus on 13 March 1781, three nights before his sister's thirty-first birthday. On the 17th he confirmed the object's motion.

189

return visit, or one of the more mysterious nebulous objects that so engaged his attention.

You, Miss Mitchell, have savoured the promise of first sighting such a possibility, and have passed your own anxious hours till the next cloudless night when you could turn your eye to the same spot of sky, your heart full of hope that your blur had not stayed put where you left it, but rather strayed among the stars, to testify by its movement, 'I am a comet, yes, and because you caught me, perhaps I may be yours!'

Dr Maskelyne was first to confirm William's find, though he declared it the oddest comet he ever did see, sans tail, sans coma, and in possession of a disturbingly well-defined disk. I think he suspected even then that William had found a planet and not a comet, which is a remarkable thing for an Astronomer Royal, truly, and the good Dr Maskelyne was not prone to creative leaps.*
Of course, the job of Astronomer Royal does not require imagination; it requires exactitude in the mapping of the heavens, at which Dr Maskelyne excelled, and yet he seemed ready and willing to jump to a new planet. Who would have guessed it of him?

It was he who urged William to write a paper for the Royal Society, which William titled simply 'Account of

* The Reverend Doctor Neville Maskelyne (1732–1811) served as England's fifth Astronomer Royal, from 1765 until his death. Acknowledging one of Miss Herschel's several comet discoveries, he called her 'My worthy sister in astronomy'.

a Comet'. One of the members of the Society read it aloud at the April meeting in London, while we remained in Bath, for William was still engaged full-time as organ master at the Octagon Chapel, and moreover the inadvertent discovery of the 'comet' had interrupted a busy programme of stellar observing and measuring separations between double stars. Though we declined to go to London, soon most of London came to us, to see our little house with its basement workshop and the seven-foot telescope in the garden.*

Shortly after William's comet aroused our interest, it took a summer holiday, spending several months out of sight in the daytime sky, so that no one could amass the observations required to establish its orbit. When it returned at the end of August, we – and here William and I were joined by I daresay half the astronomers in Europe, not to mention Russia – all fixated upon it. Night after night we strove to fit our observations along a typical comet's parabolic path, while the object refused to obey our rules, and would move stubbornly in a circular arc. All through the autumn it failed to brighten for us; it denied us the delight of seeing it flash its tail. By November the truth finally dawned: the comet was a planet at twice the distance of Saturn!

---

* The Herschels' small Georgian house, at 19 New King Street, Bath, is now open to the public as the William Herschel Museum. The seven-foot 'Uranus telescope', with its six-inch mirror, resides at the Science Museum in London.

As I have taken pains to explain to you, Miss Mitchell, our 'Eureka!' moment followed the detection of the body by more than half a year. William had uncovered one thing that proved to be quite another. When the magnitude of his feat shone clear, and word spread that he had single-handedly doubled the width of the Solar System with this distant planet, King George offered his official protection, including a handsome stipend of about two-thirds Dr Maskelyne's salary. The planet could not have arrived at a more propitious moment, given the crown's recent loss of the Colonies in America.

William, universally hailed as the first man in history to discover a planet, ceased offering music lessons and performing in concerts to become a full-time astron-omer. In France some people campaigned to name his new find 'Planet Herschel', just as yours is 'Comet Mitchell'. Men who had never before heard of William conceded that his homemade telescopes had put the tools of every great observatory to shame. Not a visitor left our home unshaken by what William had built with his own hands and at his own expense. Hardly a congratulatory letter arrived that did not include a request for William to sell the writer an instrument.

None of the adulation turned William's head, how-ever, and he would hear none of 'Planet Herschel'. We both felt that while it was all well and good for comets to carry their discoverers' names – since this practice had precedent in our field, and the number of comets might be legion – the naming of a planet, being such a much rarer occasion, begged for different criteria.

William suggested 'Georgium Sidus', to acknowledge the King's kindness, though it was quickly pointed out that any reference to national allegiance might be misplaced in a heavenly body. Many other names came forward before the idea of 'Uranus' occurred to Herr Bode in Berlin, who sought safety in mythology.* Bode published an annual ephemeris, which gave him an influential voice in such decisions, but even so, the planet answered to three names – 'Uranus' in most of Europe, 'Herschel' in France, and 'the Georgian' in England – for SIXTY YEARS before 'Uranus' gained currency. In that interim, a prodigious chemical experimenter – another German, named Klaproth – extracted a metal from pitchblende and called it 'Uranium'. He told us that alchemists of old had always given planet names to their metals, and he thought the new planet deserved to have one christened in its honour.†

The focus in astronomy remained the establishment of the planet's orbit, no matter what its name. We also wondered how 'Uranus' had escaped prior detection, for although William had sighted it through a superb telescope, other astronomers easily found it with

---

* Johann Elert Bode (1747–1826), editor of the *Berliner Astronomisches Jahrbuch*, became director of the Berlin Observatory in 1786.
† Analytical chemist Martin Heinrich Klaproth (1743–1817) isolated and named uranium in 1789.

inferior instruments once he told them where to look. This suggested that old records might yield helpful early notes on the planet's past positions, set down innocently by observers who had mistaken it for a star. Herr Bode, perhaps because of his dedication to his own yearbooks full of tables, took up this task and was soon rewarded for his effort. He found a sky chart from 1756 that included a star no longer to be seen at its given coordinates. That spot now stood empty, while the trail of the planet Uranus, as far as anyone had succeeded in describing it, would have touched that exact point in that very year. This proved a most agreeable vindication, and sent Bode scurrying to find more ancient mentions of our new planet. Indeed, William had hardly been the first to see Uranus for what it was NOT! The venerable Mr Flamsteed had listed it in his star catalogue of 1690, in the constellation of Taurus the bull.* This was not as happy a match, however, since no one could make Mr Flamsteed's star – now missing – mesh precisely with the path of Uranus as we understood it. Some were tempted to dismiss the information from Greenwich, blaming carelessness or an antiquated telescope for the discrepancy. But I was intimately acquainted with the star catalogue of Mr Flamsteed, a renowned observer in his day and the quintessential Astronomer Royal (per-

* John Flamsteed (1646–1719) became England's first Astronomer Royal in 1675, the year the Royal Observatory opened in Greenwich Park.

haps even more perfectionist by nature than his successors), and it seemed unlikely he would have erred in his notations. You can imagine the astronomical dilemma: on the one hand, we desperately needed the old data to aid our calculations, since distant Uranus moved so painfully slowly, and no one relished the thought of spending seventy or eighty years to track its motion once around the Sun! On the other hand, if the ancient observations confounded the best current sense of the orbit's shape, what help could they provide?

As our new planet continued its strange peregrinations, William built ever larger telescopes. It was through one of these, on a January night in 1787, when our thermometer registered thirteen degrees Fahrenheit, that William discovered Uranus's two moons. He never proposed proper names for either body, nor for the two satellites he found two years later at Saturn, but my nephew, who was quite the literary scholar before following his father's footsteps in astronomy (perhaps you are familiar with John's translation of the *Iliad*?), named them all. The nomenclature for the Saturn system rests squarely on the Greco-Roman myths, but John selected the Uranian lunar names from Shakespeare! As a well read librarian, Miss Mitchell, you of course recognize Oberon and Titania as the king and queen of the fairies from *A Midsummer Night's Dream*, though I believe the allusion has escaped many another astronomer.

As years passed in watching William's planet creep

From the observer's platform of his largest telescope, which boasted a tube 40 feet long and a mirror 48 inches across, Sir William Herschel (1738–1822) communicated through speaking tubes with his sister Caroline in the hut below. It took two workmen to turn the instrument on its revolving base. Through this telescope, Sir William discovered the sixth and seventh satellites of Saturn, Enceladus and Mimas, in 1789.

about the sky, the deplorable difficulties with its orbit worsened. The more we accumulated new observations, and the more we extracted ancient sightings from observatory files, the less these could be reconciled. It was proving impossible to predict where Uranus would be even a year or two hence, though most astronomers could assess the future whereabouts of Jupiter or Saturn to within a hair's breadth till the end of time. Thus the great beauty of the great Newton's contribution seemed tarnished by the recalcitrant behaviour of Uranus.

Things still stood unsettled, sad to say, when William passed away, forty years after the discovery of his planet. I left England at that point, returning to Hanover to live with my brother Dietrich. Neither of us yet realized that William's life span had equalled the 83.7-year orbital period of Uranus. (Is that not a remarkable coincidence, Miss Mitchell?!) We knew only that the errors in fitting predictions to observations were growing ever more egregious. The last explanation to reach William before he died suggested that a large comet had struck Uranus just prior to its discovery, and the impact had altered the planet's course. This supposed collision gave probable cause for the rift between the old data and the new, though it seemed almost too imaginative a solution to be believed – more like some device from Shakespeare's theatre, or the Greek tragedy, where gods descend on contraptions to tidy the loose ends of a drama.

Surely some astronomers welcomed the impact

idea, but, shortly after William's death, the orbital predictions based on a comet collision ALSO proved
incapable of mapping our planet's route. There was
nothing left for the mathematicians, I suppose, but to
insist upon another large planet lurking undiscovered
in the deep, beyond Uranus, pulling it off course. How
William would have applauded the diligence that discerned such a world in the mind's eye, before it came
to light in the sky, purely by harnessing the intellect to
paper and pencil! What would he say to the news of
the two now celebrated young gentlemen, who both
independently discovered the same new planet, without
either one's ever so much as putting his eye to a telescope, or even knowing which end of it to peer
through?*

Only think, Miss Mitchell, of the feats of computation
required to build an orbit in the air for a body not
known to exist. Think of the bewildering array of possibilities that must first be conjured and then tested one
by one to induce this hypothetical body, on its hypothetical travels, to assume responsibility for all of
Uranus's waywardness. I have heard it said M. Leverrier
covered ten thousand sheets of paper with his figuring,
and not for a moment do I doubt that estimate. Mr

* In 1845, theoreticians Urbain Jean-Joseph Leverrier (1811–1877)
and John Couch Adams (1819–1892) successfully completed their
separate calculations showing that a large, exterior planet could
account for the irregularities of Uranus's motion.

Adams cannot have done less. And yet, after such immense labour, with each man persevering unaware of the other's toil, both of them had to BEG the chief astronomers of their respective countries to point telescopes toward the heavenly vicinity where the proposed planet could be found.

That the current Astronomer Royal all but ignored the inexperienced, unpublished Mr Adams is sad, but not difficult to fathom.* M. Leverrier, on the other hand, was already famously distinguished in Parisian scientific associations, and had PUBLISHED his predicted position for the planet, yet he, too, FAILED to gain the cooperation of his national Observatory. (Were you, Miss Mitchell, perchance among the small cadre of independent astronomers who heeded M. Leverrier's call to action? I understand that several Americans tried to locate the planet according to his directions.)

The persistent M. Leverrier ultimately succeeded in circumventing official channels via his written request to young Dr Galle, a SUBORDINATE at the Berlin Observatory. Galle, fresh from his graduate studies, had had the good fortune to observe Comet Halley in 1835, and then the good sense to send his thesis to Leverrier,

* The seventh Astronomer Royal, Sir George Biddel Airy (1801–1892), is remembered for his autocratic direction of the Royal Observatory at Greenwich, and for allegedly depriving England of primacy in the discovery of Neptune.

so that an affinity had developed between them.* (I bring up these details to urge you, Miss Mitchell, always to announce your findings as soon as you possibly can, not just to garner credit for yourself when credit is due, but because our science thrives on shared information.) Galle surely knew he could lose his post for turning the telescope toward Leverrier's hunch without permission, and he must have appealed with just the right degree of earnestness and obsequiousness to Prof. Encke. Fortunately for all of them, Encke was hurrying home to his own birthday celebration that evening. Had it not been for his rushing out on account of the party preparations, he might have denied his permission.

Now picture the scene later that night, when Galle and his assistant arrive, breathless and unannounced, at Encke's house, to tell him they have actually FOUND Leverrier's planet!! Meanwhile in England, unbeknownst to all, another pair of astronomers seeks the supposed new planet in a SECRET HUNT, which has at long last been authorized by the Astronomer Royal. And where is the grand personage of the Astronomer Royal the night the new planet makes its entry on the stage of the world? Mr Airy is here in Germany(!) perhaps only a few miles from the road where Galle rushes through the dark with his extraordinary news! Why, the situation

---

* Johann Gottfried Galle (1812–1910) later succeeded Encke as observatory director, and lived long enough to witness Halley's Comet a second time, in 1910.

has all the elements of a farce, except that it embodies the finest imaginable testimony to the validity of Newton's laws.

Following on the heroic mathematics of Adams and Leverrier, and the stunning confluence of their timing, Galle's espying of the planet through the telescope could be considered almost anticlimactic. I feel certain, however, that it will make his career, and that, whatever else he accomplishes in life, Galle will be known for ever as the man who first saw Neptune – or 'Oceanus' or 'Leverrier', as its name may be, though here we are already happy with 'Neptune'.

My nephew might well have preceded Galle, and made Uranus and Neptune a pair of father-and-son discoveries(!), for in July of 1830 John's explorations had taken him to the very neighbourhood of the sky – almost to the street and house number, if you will – where Neptune was then residing, though he did not knock at that door. John's good nature kept him from expressing any personal remorse for his oversight, however, and also helped him quell this past year's awful national sparring between France and England over claim to the Neptune territory. As my nephew informs me, Mr Adams had pinpointed the planet at least ten months ahead of M. Leverrier, yet told no one except his superior at Cambridge and the Astronomer Royal at Greenwich. As a result of his cautious silence, Mr Adams is denied the laurels, and although he graciously accepts second place, his countrymen would rather see him decorated a hero. (And not a few of them would send Mr Airy to the gallows!)

Yet no rancour, I am told, divides the two key individuals, for Mr Adams and M. Leverrier established an immediate rapport when they met last June at Oxford, and grew more friendly during a stay at my nephew's home in July. I suppose the strength of their common obsession unites them. They are as drawn to each other as their planet and my brother's planet are bound by the laws of celestial mechanics. For a long time each of these men was ignorant of the other, acting independently of the other, just as Uranus and Neptune seemed unaffected by each other when parted by the vast separations which their orbits allow. But soon after my brother discovered Uranus, his planet neared the environs of Neptune, where the two bodies – one in the limelight, the other behind the scenes – together revealed the full force of their mutual attraction.

With hindsight, it is easy to understand why Uranus began accelerating at an ever-growing rate from about the time of its discovery in 1781 up until it reached conjunction with the UNSEEN and much SLOWER Neptune in 1822. After Uranus overtook Neptune in that year (the same year that death overtook my William), the gradual deceleration commenced, and precipitated the crisis in prediction that brought Adams and Leverrier, each for his own reasons, to consider THE PROBLEM WITH URANUS, which they proved to be THE EXISTENCE OF NEPTUNE.

Earlier I remarked how the number of William's years compared to the period of his planet; surely the slow going of Neptune will exceed the lifetimes of Adams and Leverrier combined, and perhaps that of Galle as well.*

And now the newfound moon at Neptune obliges our two champion calculators in their continuing considerations. How quickly this body has stepped forward, as though to offer itself as the perfect vehicle for refining their necessarily rough estimates of Neptune's mass.† Neither Adams nor Leverrier could help but overestimate the mass of the hypothetical Neptune, since they both overestimated its distance from the Sun, but, given the way bodies balance mass against distance under the law of gravity, all's well that ends well, and the smaller, closer Neptune can wield as much power in reality as ever the larger, further one did on paper. The new understanding reveals Neptune to be the twin brother of Uranus, at least insofar as their mass is concerned.

How long do you suppose it may take to uncover more facts of their planetary lives, Miss Mitchell? When

* Neptune takes 164 years to complete a single orbit – longer than the 66 years of Leverrier's life added to the 73 of Adams's, but Galle's 98 years tip the balance.
† Within weeks of Neptune's first sighting (on 23 September 1846), amateur astronomer William Lassell (1799–1880) of Liverpool discovered its largest moon, Triton, on 10 October. Other astronomers confirmed the find the following July.

will we say what metals they churn, what gases they breathe? No doubt forthcoming discoveries in astronomy will require ever larger, ever more powerful telescopes. Even if we are to see brilliant minds intuit the locations of new planets on the strength of theory and calculation alone, will we not need great tools to pry those deduced worlds from the realm of the invisible? The largest of William's reflectors reached forty feet of length, with a mirror four feet in diameter, but the huge mirror became so often tarnished that William abandoned it for a smaller, more serviceable instrument. My nephew formally decommissioned the forty-foot behemoth a few years ago at Christmas, when he gathered with Margaret and all their children inside the tube, to sing a ballad he composed for the occasion. But I predict clever artificers will soon step forward, perhaps in your lifetime, and devise larger and bolder new designs capable of reaching far beyond the limit of William's daring, to collect vast oceans of light from space.

In anticipation of what we may jointly see, and again with profoundest heartfelt congratulations, I remain,

Yours most sincerely,

Caroline Lucretia Herschel.

POSTSCRIPT

Miss Herschel and her brother always maintained the discovery of Uranus had been no lucky accident, but the fruit of long years spent building a superior instrument and constantly practising upon it.

'To make a person see with such a power,' Sir William wrote, 'is nearly the same as if I were asked to make him play one of Handel's fugues upon the organ.'

When the planet's rings turned up unexpectedly two centuries later, that discovery, too, was labelled accidental. But it had taken ten astronomers, packed into the cargo bay of an airborne observatory flying over the Indian Ocean, bent on assessing the exact dimensions of Uranus by chasing its predicted passage in front of a star, to encounter such surprises by such luck.

About half an hour prior to Uranus's anticipated eclipse of the star, on 10 March 1977, the star momentarily winked. It winked again, several more times, until Uranus completely obscured its light for twenty-two minutes. After the star re-emerged from behind the planet's disk, it resumed its winking to repeat the same on–off pattern in reverse, as though it had encountered mirror-image obstacles on the other side. Astonished, the astronomers spoke excitedly among themselves of a possibly ringed Uranus even before their historic flight landed, although caution and disbelief delayed their public announcement of the rings for several days.

Sir William himself had once reported seeing a ring at the planet he discovered, but later retracted the claim as a mistaken perception. He could not possibly have spied Uranus's ultra-dark, ultra-thin hoops of tightly packed icy rock and dust, even with the best of his excellent telescopes, for the rings reflect too little visible light. They disclosed their presence only by blocking the light of a star, and they remained nine invisible shadows over the ensuing decade, until they could be visited and imaged close-hand.

The rings naturally circle the planet's widest part, at the equator. But Uranus, having been knocked over aeons ago by the forceful blow of an extremely large impactor, reclines on its equator. As a result, its rings don't encircle the planet horizontally, the way Saturn's do, but stand upright, giving ringed Uranus the semblance of a bull's-eye target hung on the sky. Through this target, like an arrow on a near-miss trajectory, shot the *Voyager 2* spacecraft on its January 1986 flyby of Uranus.

The spacecraft discovered two additional faint rings around Uranus and ten tiny satellites. Astronomers had predicted a large cast of small moons to support the Uranian rings' sharp borders, and the sudden bounty of real bodies forced them to brush up their Shakespeare. Cordelia, Juliet, Ophelia, Desdemona and the like thus joined company with Titania, Oberon and three other previously known moons. Since 1992, advanced Earth-based and Earth-orbiting telescopes have ferreted out still more minor satellites, duly named for Shakespearean magicians, monsters and minor characters.

Most of these moons appear as dark as the rings, as though coated with soot. Perhaps the same collision that upended Uranus long ago shocked the chemistry of its carbon-containing compounds, raising enough black dust to sully all the planet's companions.

In contrast to the dingy moons and rings, Uranus itself appears a pale blue-green pearl, light and luminescent. Its near twin, Neptune, reveals a more complex beauty in subtle stripes and spots of royal to navy blue, azure, turquoise and aquamarine. Both planets frost their upper atmospheres with frozen crystals of methane, which absorb the red wavelengths from incoming Sunlight, and bounce the blues and greens back into space.

Under those bluish hydrogen-helium skies, neither Uranus nor Neptune knows any solid surface. Instead, their atmospheric gases give way to interior gases that progressively thicken and compress under the mounting pressures at deeper levels, and terminate at the planets' rock-ice cores.

Uranus and Neptune constitute their own class of Solar System objects – the 'ice giants'. Each one vastly exceeds the mass of Earth (Uranus by a factor of fifteen, Neptune seventeen), yet both are dwarfed in turn by the 'gas giants', Jupiter (318 Earth masses) and Saturn (ninety-five). The ice giants might have reached their own greater proportions, if only they hadn't stood in line behind the gas giants at the feast of planetary accretion.

The 'ices' that characterize the deep atmospheres of Uranus and Neptune comprise water, ammonia and methane. Planetary scientists call these compounds ices because they solidify at cold temperatures. Pressure-cooked inside Uranus and Neptune, the ices no doubt boil as oceans of water-ammonia-methane broth. The hot soup still counts as 'ice' in the parlance of planetary science, however, like the 'hot ice and wondrous strange snow' of *A Midsummer Night's Dream*.

Within the liquid turmoil of these planetary mantles, where boiling ices mix with bits of molten rock, the turning of Uranus and Neptune bestirs electric currents that generate global magnetic fields around both worlds.

Uranus and Neptune spin at similar rotation rates (seventeen hours and sixteen hours, respectively), but their days pass in nothing like similar fashion, because the unusual prone posture of Uranus confounds the meaning of days as seasons change. Lying on its side, and taking nearly eighty-four Earth years to complete a single revolution, Uranus spends twenty years of each orbit with its south pole facing Sunward, and later another twenty years with its north pole toward the Sun. At such times, the planet's rapid rotation fails to produce a cycle of light and darkness, so the 'days' (and 'nights') last a full two decades. During the two twenty-year periods when the Sun strikes Uranus on the equator, however, the days dwindle to about eight hours, followed by nights of equal length.

Neptune's 29-degree tilt – about on a par with Earth, Mars, or Saturn – keeps days of more consistent length all through its inordinately long years, each the equivalent of 163.7 Earth years, or nearly double those of Uranus.

Very little light or warmth from the Sun reaches across the two billion miles to Uranus, and even less arrives at Neptune, another billion miles further away. Yet the high atmospheres of both planets register the same low temperature, and this likeness exposes an important difference between them: the more distant Neptune generates considerably more heat from within.

Neptune's heat powers active weather patterns, with dark storms and white clouds borne across the planet's blue expanses on swift winds. Some such tempests resemble the size and shape of Jupiter's Great Red Spot, though they seem to freely change shape as they swirl. They also roam from one latitude to another, dissipating as they go, instead of persisting confined in any specific zone.

Before *Voyager 2* flew by Neptune in 1989, the planet had just two known moons. The larger one, first observed by William Lassell in 1846 and later named Triton (for Neptune's sea-god son), amazed its discoverer by orbiting the planet *backwards*. Neptune probably captured this moon – a body the size of the planet Pluto – and forced it into orbital submission. The second moon, Nereid (a sea nymph), was discovered and named by Gerard Kuiper in 1949.*

Voyager 2 found six small, dark satellites orbiting near and among Neptune's dim, dusty, icy rings. These moons – Naiad, Thalassa, Despina, Galatea, Larissa and Proteus (all named for sea deities) – cause the ring particles to bunch up in disorderly clumps. From a distance, silhouetted against the backdrop of stars, the rings create the illusion of fragmentary arcs because they block starlight on one side of Neptune or the other, but not both. Only on close inspection do the partial curves link up, along thin connecting bridges of material, into complete rings.

Although no spacecraft has visited either ice-giant planet since the 1980s, the pace of discovery at Uranus and Neptune has picked up of late, thanks to observations made from Earth and near-Earth via infrared radiation – the very region of the electromagnetic spectrum that Sir William Herschel discovered in 1800.

Experimenting with thermometers and a prism, Sir William had taken the temperature of Sunlight's various colours, noting how the mercury rose from violet through to red, and *continued* rising in what he called

* Dutch-American astronomer Gerard Peter Kuiper (1905–1973) is generally considered the father of modern planetary science.

the 'invisible light' or 'calorific rays' beyond the red. But he never could apply this important discovery to his own astronomical researches, because water vapour in Earth's atmosphere – the same feared enemy that made Sir William rub his skin with an onion to ward off the ague while he braved the damp night air – blocks out most infrared emissions from planets and stars.

Orbiting telescopes, however, transcend the interference of atmospheric moisture. From a high perch, 375 miles above Earth's surface, the Hubble Space Telescope's infrared camera has followed recent changes on the ice giants. Large ground-based telescopes, too, specially equipped and set at high altitude in Hawaii and Chile, can now collect and amplify the few wavelengths of infrared radiation that do penetrate Earth's atmosphere. Detailed new time-lapse pictures show a dark hood spreading over bland Uranus's south pole, as summer slowly draws to a close there, while large bright clouds gather in the northern hemisphere. As the planet progresses to a new season, it turns its thin rings to face Earth edge-on. (Had they not already been discovered in 1977, the rings would avoid discovery now.) On Neptune, the current build-up of bright new clouds over the southern hemisphere progressively lightens the colour of the sky.

The planet Neptune, fished from the pool of space as the answer to a dynamical puzzle, repaid the favour of its discovery by posing a new dynamical problem. Early

211

in the twentieth century, the conviction that Neptune alone could not account for all of Uranus's orbital vagaries (not to mention a few vagaries of Neptune's own) fomented a 'Search for Planet X', which culminated in the discovery of Pluto.* Recent recalculations, however, prove the mass of Neptune to be sufficient after all. *Voyager 2*, the only spacecraft to visit Jupiter, Saturn, Uranus and Neptune, provided precise measurements of the pull that each giant planet exerted on the craft's own small body. These results forced an upward revision of the mass estimate for Neptune, amounting to one-half of 1 per cent, or just enough to render Pluto irrelevant in shaping Uranus's orbit. As in Miss Herschel's time, the wanderings of Uranus can still be laid to the presence of Neptune.

But if Uranus begs no further explanation from the Solar System's outer limits, the moons of Neptune do. The strange orbital patterns of Triton and Nereid point accusingly to origins in the outer depths. Out there, far beyond the precinct of the major planets, and lying just below the current threshold of detection, untold numbers of objects still await discovery.

---

* Pluto was discovered by American astronomer Clyde W. Tombaugh (1906–1997), and the finding made public on 13 March 1930.

11

UFO

My Grandpa Dave, a teenaged alien, arrived at Ellis Island alone in a crowd, then worked a private aeon of man-hours – sewing buttonholes, delivering soda water – to bring his mother, father and younger brothers light-years over the ocean in his wake. 'Mama!' he screamed to her across the packed immigration hall, where health officers had detained her for an eye infection more foreign and unwanted than she. Deportation seemed imminent, but the officials, moved by the emotion of the mother-and-son reunion, instead welcomed Malka Gruber to America.

My mother could never tell this story without crying, as though she had witnessed its embraces, or suffered the threatened exclusion. Even when she became a very old woman, the retelling of that moment long before her own birth would catch in her throat. I, too, yet another generation removed, can turn weepy over it

– an empathic response that predisposes me, a recent psychological study has shown, to the creation of false memories, such as the recollection, now prevalent among an estimated three million Americans, of having interacted with visitors from another planet.

The idea that aliens might hail from other planets – as opposed to the 'old country' my grandparents and other immigrants left behind – gained credence in 1896. In that year Percival Lowell, scion of the wealthy and privileged Boston Lowells, woke the public to the plight of pitiable Martians who had all but exhausted their global water supply and were husbanding what remained via canals crisscrossing their world.

Lowell had spent his early manhood travelling in Europe, the Middle East and the Far East, proving his facility with language and flair for explaining foreign ways to Yankees. In preparation for the 1894 close approach of Mars, Lowell indulged his passion for astronomy by establishing a private observatory in Flagstaff, Arizona, free from the control of any academic, military, or government authority. The thirty-nine-year-old Lowell so overextended himself – building, staffing and equipping the site on 'Mars Hill', then observing the planet from May of 1894 to April 1895, collecting his thoughts and nine hundred drawings into his popular book, *Mars*, and addressing numerous general audiences during a lengthy lecture tour before hastening to Mexico in 1897 to catch the next Mars opposition – that he collapsed. Lowell's attack, diag-nosed as 'severe nervous exhaustion', disabled him for four years.

When he returned to Flagstaff from Boston in 1901, he found his staff demoralized by the fuss over the canals. Lowell's sensational conclusions and rush to publication had made Mars Hill a laughing stock among professional astronomers. Although immune to criti-cism himself, Lowell, who had excelled in mathematics at Harvard, determined to restore his observatory's reputation by calculating the whereabouts of a ninth planet. Enough discrepancy still disturbed the orbit of Uranus to suggest that the spectacular feats of Adams and Leverrier in the previous century could be repeated, on American soil, to yield a new world beyond Neptune.

Lowell called his quarry 'Planet X'. He pursued it enthusiastically, albeit unsuccessfully, until his death in 1916. For the next ten years, Lowell's widow hamstrung all observatory operations by disputing the intent of his will. The planet search finally resumed in 1929, with a new dedicated telescope situated in a new dome on Mars Hill, and a raw youth – an amateur with only a high school education – hired through the mail to man it.

Clyde Tombaugh, perhaps the most upstanding, hardworking, unimpeachably decent young man ever to leave the wheat fields of Kansas for the astronomical

217

high ground of Arizona, traded his life's savings for a one-way train ticket to Flagstaff. On an impulse, he had sent his drawings of Jupiter and Mars, as seen through his own homemade telescope, to the Lowell Observatory. The director, impressed, had written back to him, inquired after his health, then offered him the difficult, low-paying job of methodically searching the heavens, inch by inch.

In comparison to Johann Galle, who homed in on Neptune after only an hour's guided effort, Clyde Tombaugh spent ten months of cold nights in the open dome on Mars Hill, photographing the sky in a meticulous series of hours-long exposures. After he processed the plates, he examined and compared them by pairs with a microscope, scrutinizing the many thousand points of light to see which, if any, had shifted position from one image to the next. Through this tedious process, he located Lowell's Planet X in mid-February of 1930. The planet was travelling among the stars of Gemini, at a rate that suggested it lay a billion miles past the orbit of Neptune – and just about at the coordinates Lowell had predicted.

Tombaugh's cautious older colleagues made him confirm and re-confirm his discovery for three weeks before they released an official announcement, with all due protocol, by mailing a detailed circular to every observatory and astronomy department they could name. The world went wild. The Associated Press carried the news by wire, and when the story reached *The Tiller and Toiler*, the weekly paper of Pawnee County, Kansas, the editor phoned Muron and Adella Tombaugh on their farm in Burdett to ask, 'Did you know your son discovered a planet?'

Clyde was twenty-four years old. Having made history, he took a leave of absence from the Observatory to attend Kansas University and earn his degree in astronomy.

A spate of telegrams hit Flagstaff in response to the Pluto news, followed by sacks of mail and soon hundreds of visitors every day. Reporters clamoured for photos, but the discovery images no doubt disappointed most expectations. They looked like a pair of ink spatters, differing from each other by the placement of a single spot no bigger than the dot of an 'i'.

The best available instruments strained for better views of Pluto, but few could resolve the dim dot into a planet-like disk, let alone discern features on its surface. Indeed, Pluto is so small and so far away that even today, the most detailed portraits obtained with the Hubble Space Telescope reveal merely a bleary sphere in shades of grey, as unsatisfying and lacking in detail as a faked photo of a UFO.

Doubting astronomers in 1930 challenged the claim that Lowell's Planet X had been found. *That* planet had promised to exceed the mass of Earth several times over, to be big enough to sway Uranus and Neptune. The newly discovered planet, however, seemed much too insubstantial to tug giants.

Since the 1930s, Pluto has shrunk ever smaller with each new improvement in measuring techniques. Its mass dropped from the original estimate of *ten times* the mass of Earth to *one-tenth* the Earth's mass, to *one-hundredth*, to about *two-thousandths*. Meanwhile Pluto's diameter diminished from an Earthlike 8,000 miles to a mere 1,500 at most. Pluto turns out to be smaller than the planet Mercury, and smaller also than seven Solar System satellites, including Earth's Moon. Pluto's own moon, Charon, discovered in 1978, measures half the width of Pluto itself, while most other moons' diameters are only one-hundredth that of their parent planets.

Pluto's precipitous size decline over the fifty years following its discovery prompted two planetary astronomers to publish a whimsical graph in 1980, depicting the diminution of Pluto as a function of time, and predicting the planet would soon disappear!

Shrivelled and ridiculed, Pluto was altogether stripped of its reason for being after *Voyager 2* passed Neptune in 1989. The need for a ninth planet vanished then in the realization that Neptune and Uranus balanced each other's orbital anomalies. The calculations that had led Lowell to the prediction of Planet X apparently held no more water than his Martian canals. Pluto had entered into popular awareness as the answer to a meaningless question.

In 1992, a small new Pluto-like body turned up on the fringes of the Solar System, followed in 1993 by another five like it, and over the next few years by several *hundred* more. This outlying population offered Pluto a new identity – if not the last planet, then the first citizen of a distant teeming shore.

Pluto seemed to be reliving the history of the first asteroid, Ceres. Hunted, like Pluto, on mathematical grounds, Ceres was greeted as the 'missing planet' between Mars and Jupiter at the start of the nineteenth century. When continued observation proved Ceres too puny and its type too numerous to rank with the major worlds, astronomers reclassified the lot as 'asteroids' in 1802, and later as 'minor planets'.

No public outcry attended the application of those lesser terms to Ceres, Pallas and their companions. Pluto, in contrast, retains an emotional hold on planethood. People love Pluto. Children identify with its smallness. Adults relate to its inadequacy, its marginal existence as a misfit. Anyone accustomed to a quota of nine planets – anyone averse to changes in the status quo – baulks at disqualifying Pluto on a technicality.

Even within the six-hundred-member fraternity of planetary astronomers, opinions on Pluto stand angrily divided. Is it a planet or isn't it? Unfortunately, the word 'planet', coined long before science demanded much specificity of definition, cannot support the many possible gradations of meaning implied by recent discoveries.*

The campaign to drop Pluto from the planet registry, although widely perceived as a shameful demotion, in fact salutes the greater diversity of an expanded Solar System. Pluto and its ilk fill a doughnut-shaped 'third zone' that extends outward from Neptune to at least fifty times the Earth–Sun distance. Since all the objects in this territory differ fundamentally from the terrestrial worlds in the first zone or the gas and ice giants in the second, they have been given their own new designation of 'ice dwarfs', or 'Kuiper Belt objects (KBOs)'.

The eponymous Gerard Kuiper first conceived of these bodies in 1950. Born and educated in the Netherlands, Kuiper emigrated to the United States in 1933 and became the country's major proponent of planetary studies, with discoveries ranging from the atmosphere of Saturn's largest moon, Titan, to new satellites for Uranus and Neptune. Looking to the future, Kuiper forecast that Pluto, the lone outcast of the Solar System, would be found to have hundreds or thousands of fellow travellers. Half a century later, when Kuiper's myriads began to materialize in the trans-Neptunian deep, astronomers recognized them as his hypothesis come real.

* Like 'planet', the word 'life' poses similar difficulties for astrobiologists: a wildfire, for example, exhibits lifelike behaviour as it takes in oxygen, grows, moves, consumes, even generates new fires with its own sparks, but it is not 'alive'.

The steadily increasing census of the Kuiper Belt counts Quaoar, Varuna and Ixion, all discovered in 2001 and 2002, among its larger constituents. Their names reflect a modern ethic of ethnic awareness: Quaoar is the creation force recognized by the Tonga tribe, the original inhabitants of what is now Los Angeles.

Pluto, the premier object in the Kuiper Belt, follows a steeply inclined and highly elliptical orbit. Over a period of 248 years, Pluto alternately soars above the plane of the Solar System and dives below it, strays out to almost twice Neptune's distance from the Sun at one extreme and ducks inside the orbit of Neptune at the other.* This wandering path, so different from that of any other planet, helped brand Pluto as an oddball from its earliest days. By the standards of the Kuiper Belt, however, the orbit appears common. Some 150 other Kuiper Belt objects trace the same course, and they all avoid collision with Neptune thanks to the resonance agreement among them: Neptune circles the Sun three times in the time it takes Pluto and company to go around twice. When Pluto trespasses into Neptune's orbit, it does so always at the height of its swing, leaving Neptune far below and at least a quarter-turn away.

---

* Pluto last dipped within the orbit of Neptune in 1979 and emerged in 1999. At perihelion, in 1989, Pluto lay approximately one billion miles closer to Earth than it had been when discovered in 1930.

Pluto spins round its own axis once every six days, rotating the dim splotches of its vague landscape in and out of view. Like Uranus, Pluto lies on its side, the victim of a prior collision. Indeed, planetary scientists believe that a single impactor knocked down Pluto and chipped off its moon Charon in one blow.

Pluto and Charon, only about 12,000 miles apart, lock each other in orbit around a point partway between them. They both rotate at the same pace while jointly circling this point, so that each keeps the same face always turned towards the other. The uniqueness of their orbital engagement has recast Pluto and Charon as 'Pluto–Charon', the first known example of a true 'double' or 'binary planet'.

Less than a decade after Charon's discovery, Pluto and Charon oriented themselves in space so as to take turns eclipsing one another, as viewed from Earth. Such a fortuitous arrangement can occur only twice during Pluto's orbit, or once every 124 years. Beginning in 1985, astronomers took advantage of the numerous mutual occultations to derive the best possible approximations of the two bodies' mass, diameter and density. At about twice the density of water, both Pluto and Charon are more dense than any of their gaseous giant neighbours, though not half so dense as the iron-rich terrestrial planets Mercury, Venus and Earth.

Perhaps two-thirds to three-quarters of Pluto consists of rock, and the rest ice. Above Pluto's bedrock of water ice, patches of frozen nitrogen, methane and carbon monoxide have been identified from afar. When Pluto warms itself inside Neptune's orbit for two decades every two centuries during its nearest approach to the Sun, ices on the planet's surface partially evaporate to form a puffy, rarefied atmosphere. Later, as Pluto recedes from the Sun and its temperature drops back to a frigid normal (about two hundred below zero centigrade), the atmosphere falls down and coats the ground, especially around the poles, with fresh, exotic snow. In this regard Pluto behaves somewhat like a comet (which would also heat up and blow off icy gas upon nearing the Sun), though it remains too distant to create any great display.

By the time the Sun's light reaches Pluto, distance has dimmed it a thousand-fold, so that the Sunlit planet in daytime resembles a winter evening by Moonlight. On Pluto's reflective landscape, bright surface frosts coexist with dark areas that may represent rock outcrops or deposits of organic compounds extorted from the ice by the Sun's ultraviolet light. Polymers in carbon-rich colours – pink, red, orange, black – probably proliferate on Pluto.

Despite the Pluto–Charon similarity in composition and the pair's shared common origin, the moon's smaller mass and lower gravity cause it to lose its grip on gases. Molecules vaporized from Charon's surface do not hover aboveground waiting to return as snowflakes;

they simply escape into space. As a result, Charon reflects considerably less light than Pluto, and its surface will most likely appear dull-neutral in photographs when the binary worlds of Pluto–Charon are eventually visualized by a visiting spacecraft.

All past attempts to mount a mission to Pluto failed at the funding stage – before any craft could reach the launch pad, much less begin the long journey. Now, after the disappointing cancellations of projects such as 'Pluto Express' and 'Pluto Fast Flyby', Plutophiles finally have a scout being readied for the Kuiper Belt. NASA's minimalist 'New Horizons', equipped to map and image Pluto, Charon and at least one other KBO at close range, should see its promised lands in 2015. By then, the number of known KBOs may have increased exponentially, from the eight hundred identified to date, to the hundreds of thousands more anticipated.

Already the demographics of the Kuiper Belt hint at great waves of migration that characterized early Solar System history. All the KBOs, it seems, were exiled to their present locations, from positions closer to the Sun, at the time the giant planets were completing their own accretion. Jupiter and Saturn swallowed some small planetesimals in their vicinity and accelerated many more with such force that the bodies were banished from the Solar System. While Uranus and Neptune also participated in this planetesimal diaspora, they lacked the power to hurl objects entirely beyond the Sun's reach, and relegated them instead to the Kuiper Belt.

As a result of these displacements, Jupiter lost some of its orbital energy and moved in closer to the Sun. Saturn, Uranus and Neptune, in contrast, gained energy and edged further away. Pluto, which is thought to have occupied a round, regular orbit at this early stage, was shoved outward by the gravitational influence of Neptune. Over tens of millions of years, Neptune forced Pluto, the ultimate expatriate, to follow an ever more tilted, more elliptical course.

Pluto and the other Kuiper Belt residents have thus been worked over by events in the Solar System. Although scientists had hoped the Kuiper Belt might preserve pristine material, unchanged since the formation of the Sun, they now see it as a war zone where bodies have been deposited and left to fray each other. The true, untainted genealogical roots of the solar family must be pursued at a still further remove.

Today, ever more distant worldlets are swimming into view beyond the Kuiper Belt. The planetoid Sedna, discovered in 2003 and named for the Inuit goddess of the icy sea, is currently the coldest, most distant known member of the Solar System. About half the size of Earth's Moon, Sedna seems to ply an orbit that reaches to nine hundred times Earth's distance from the Sun, and that takes ten thousand years to complete.

Further on, between the dim body of Sedna and the

bright spectacle of the distant stars, astronomers expect to encounter a spherical swarm of trillions more small objects surrounding the Solar System. Among these frozen leftovers of creation lie perhaps the profoundest answers to the question of where we came from.

The outlying ancient debris distributes itself over such a distended area that the Solar System's periphery is transparent as a crystal ball. Through the bubble of its outer boundary we can see for ever – across the Milky Way home of our Sun, into the other galaxies that twirl like pinwheels strewn across the Universe, their many billion stars frothing with planets.

Sometimes the stupefying view into deep space can send me burrowing like a small animal into the warm safety of Earth's nest. But just as often I feel the Universe pull me by the heart, offering, in all its other Earths elsewhere, some larger community to belong to.

12

PLANETEERS

There was a big party at Andy Ingersoll's house in Pasadena the night after the *Cassini* spacecraft flawlessly inserted itself into orbit around Saturn in the summer of 2004. The music and dancing, the food and drink, the camaraderie were really intended for the scientists and engineers whose years of work had led to such happy cause for celebration, but a few outsiders standing in the right place at propitious moments had also been invited.

When I arrived, too early, I found our host, a senior and much venerated planetary scientist at the nearby Jet Propulsion Laboratory, fabricating a model Saturn to hang at the driveway as a location marker for the two hundred-odd guests. He had an old red tether ball with the cord still attached to it, and there on the cleared kitchen table he was cutting poster-board rings of the proper proportion to tape in place around it. A colleague came in through a back door and casually began offering technical advice, as though the current prank were some new research challenge. Within minutes, they had Saturn on a string, dangling from a branch.

Ingersoll, tall and bony, excels at modelling planetary atmospheres. He works the data points collected by telescopes and spacecraft – temperature readings, gas abundances, fluid pressures, wind speeds, cloud patterns – into sophisticated weather analyses. His journal publications have titles such as 'The runaway greenhouse: A history of water on Venus', 'Dynamics of Jupiter's cloud bands' and 'Seasonal buffering of atmospheric pressure on Mars'. He could likely match wits with any of the most famous astronomers in history, but he is unlikely to dominate the future, the way a Cassini or a Huygens persists today, because the nature of science itself has changed, from a field for lone geniuses to a collaborative effort.

The ebullient early-bird volleyball game in the Ingersolls' backyard ended about half an hour later, when caterers came to lay out the long buffet and set up tables and folding chairs around and under the trees. In the group I happened to sit with, half the people were speaking Italian and the other half a British-accented English. The party grew steadily more multinational because the *Cassini* spacecraft is global in every way. As the joint project of NASA, ESA (European Space

Agency), and ASI (Agenzia Spaziale Italiana), *Cassini* represents seventeen countries and the pooled talents of some five thousand individuals, including a team of seamstresses who custom-tailored the spacecraft's golden lamé thermal suit, to protect its instruments from dust-sized micrometeoroids and the extreme cold of the near-Saturn environment.

Each wave of latecomers to the party brought fresh bulletins from the lab. Some of the guests hadn't slept for days, and looked it, but they relished the cause of their exhaustion. The news from *Cassini*, chattering into the Deep Space Network's receivers in Spain, Australia and California, was all good. Ideal, in fact. The craft's first close-up pictures of Saturn's rings exposed such depth of exquisite detail that one astronomer had accused another, further up in the data stream, of doctoring the images as a practical joke.

The adrenaline rush that most of these men and women had experienced the previous evening during *Cassini*'s passage through Saturn's rings now mellowed into a general euphoria, a veritable Saturnalia. As the revellers toasted the present success, they also hailed the mission's next major phase – the delivery, six months hence, of *Cassini*'s robotic passenger, the *Huygens* probe, to Saturn's largest moon, Titan. That grand satellite, a body bigger than Mercury or Pluto, and possessed of a thick orange atmosphere as rich in nitrogen as our own air, had long intrigued scientists for its promised insights into conditions on the early Earth before life began. No one yet knew what lay on the smog-obscured surface of Titan, but many scientists were willing to wager great lakes filled with chill liquid methane and other hydrocarbons.

'I dream of landing in an ocean,' *Huygens* project scientist Jean-Pierre Lebreton had said the day before the party at a press briefing. 'To go to Titan now is like going back in time to Earth four billion years ago.'

From the moment Christiaan Huygens first saw Titan from The Hague in 1655, he called it simply 'Saturn's moon'. Jean Dominique Cassini, who found four other Saturnian moons between 1672 and 1684, was content to refer to them by number. And when Sir William Herschel sighted the *next* two in 1789, he, too, applied numerical designations. But Sir William's son, Sir John Herschel, chose names for them all from Greek mythology, beginning with 'Titan', an ancient race of giants, the youngest of whom was Saturn.*

In December 2004, on schedule, *Cassini* released the *Huygens* probe it had carried on its seven-year journey

* Later astronomers followed suit up to Pan, the eighteenth satellite of Saturn, discovered in 1990. The next twelve moons, including Mundilfari and Ymir, received names from broader cultural contexts, while a few new ones detected by *Cassini* still go by preliminary designations, such as S/2005 S1.

from Cape Canaveral, and nudged it toward Titan. For the next three weeks *Huygens*, still asleep, obediently coasted to its rendezvous, while *Cassini* executed another long loop around Saturn and returned in time for the planned excitement.

On 14 January 2005, *Huygens*'s internal alarm woke its systems to prepare for action at Titan. The probe hit the atmosphere with its heat shield forward, decelerated in the friction of the thick air, and parachuted to a perfect landing. It sampled the clouds and haze all during its two-and-a-half-hour descent, and when it got close enough to the moon's frigid surface (about thirty miles, as measured by the onboard radar) it photographed that, too, then relayed its findings to *Cassini*, and *Cassini* forwarded them to Earth.

On Titan, *Huygens* saw sights as familiar as clouds changing shape, as strange as the novel landscapes of an alien world, too unusual to be parsed.

The fact that *Huygens* survived touchdown and continued to broadcast evidence of its own robust health for several hours upset the widespread expectation of its drowning in a methane sea. However, the great dark expanse where *Huygens* laid itself to rest, now called Xanadu, should not be viewed as the site of a failed prediction, but rather the embarkation point for another new way of imagining the content of the Solar System, and of other solar systems as well.

I wish I could tell you all that happened next, how

the interpretation of the *Huygens* data played out, what *Cassini* encountered as it swept by this or that Saturnian satellite – Mimas, Enceladus, Tethys, Dione, Rhea, Iapetus – on the busy itinerary of its ongoing exploration. But what book can keep abreast of current events in an active field of study? If reading these pages has helped someone befriend the planets, recognizing in them the stalwarts of centuries of popular culture and the inspiration for much high-minded human endeavour, then I have accomplished what I set out to do.

For myself, I confess that none of the truly staggering data I have been privileged to share here has altered the planets' fundamental appeal to me as an assortment of magic beans or precious gems contained in their own cabinet of wonder – portable, evocative and swirled in beauty.

ACKNOWLEDGMENTS

Thank you to all the scientists and advisors who gave me such generous portions of their time, or enthusiasm, or both: Diane Ackerman, Kaare Aksnes, Claudia Alexander, Mara Alper, Will Andrewes, Victoria Barnsley, Jim Bell, Bob Berman, Rick Binzel, William Brewer, Joseph Burns, Donald Campbell, John Casani, Clark Chapman, K. C. Cole, Guy Consolmagno, Lynette Cook, Kathryn Court, Dave Crisp, Jeff Cuzzi, David Douglas, Frank Drake, Jim Elliot, Larry Esposito, Tony Fantozzi, Timothy Ferris, Jeffrey Frank, Lou Friedman, Maressa Gershowitz, George Gibson, Owen Gingerich, Tommy Gold (died 2004), Dan Goldin, Peter Goldreich, Donald Goldsmith, David Grinspoon, Heidi Hammel, Fred Hess, Susan Hobson, Ludger Ikas, Torrence Johnson, Isaac and Zoe Klein, E. C. Krupp, Nathania and Orin Kurtz, Barbara Lebkeucher, Sanjay Limaye, Jack Lissauer, Rosaly Lopez, M. G. Lord, Stephen Maran, Melissa McGrath, Ellis Miner, Philip Morrison (died 2005), Michael Mumma, Bruce Murray, Keith Noll, Doug Offenhartz, Donald Olson, Jay Pasachoff, Nicholas Pearson, Elaine Peterson, David Pieri, Carolyn Porco, Christopher Potter, Byron Preiss, Pilar Queen, Kate Rubin, Vera Rubin, Carl Sagan (died 1996), Lydia Salant, Carolyn Scherr, Steven Soter, Steve Squyres, Rob Staehle, Alan Stern, Dick Teresi, Rich Terrile, Peter Thomas, John Trauger, Scott Tremaine, Alfonso Triggiani, Neil deGrasse Tyson, Joseph Veverka, Stacy Weinstein, Joy Wulke, Paolo Zaninoni and Wendy Zomparelli.

Two people truly championed this project and guided it to its present form: Michael Carlisle of InkWell Management, my wonderful agent, by wanting to know the difference between the Solar System and the Milky Way, and likewise between the galaxy and the universe; and Jane von Mehren, former editor-in-chief and associate publisher at Penguin Books, who responded to my manuscript with dozens of astute questions and hundreds of helpful suggestions, all tendered with patience and wisdom. Michael and Jane would not have considered themselves 'planeteers' at the outset, but now that we have made this journey together, they both look to the night sky much more often than before.

GLOSSARY

APOGEE   the greatest distance from Earth reached by the Moon in its monthly orbit, or by an artificial satellite circling our planet.

APPARENT MAGNITUDE   the brightness of a heavenly body as viewed from the vantage point of Earth, expressed as a number; the lower the number, the brighter the object appears. (The Sun, with an apparent magnitude of −27, is the brightest object in Earth's sky, though if it were judged according to its *intrinsic* brightness, or absolute magnitude, it would pale in comparison to larger stars.)

AREOGRAPHER   one who makes maps of Mars (Ares).

ASTEROID   a minor planet, generally small and rocky, some one hundred thousand of which orbit the Sun in the wide gap between Mars and Jupiter.

CARTOUCHE   in cartography, a decorative emblem bearing text such as the title of the map, or the scale, and often including symbols of the regions represented.

COMA   the fuzzy envelope surrounding the nucleus of a comet.

COMET   a small icy body orbiting the Sun in a highly elliptical orbit, changing its appearance on close solar approach by emitting jets of gas and dust.

CORONAE (singular CORONA)   sets of concentric rings surrounding features such as domes and depressions, unique to Venus, found where her surface crust is thinnest.

DURICRUST   loosely cemented dust seen on the surface of Mars, thought to be formed by the deposition and evaporation of water and carbon dioxide.

ECCENTRICITY   the degree to which a body's orbit deviates from a circle. (The orbit of Pluto is highly eccentric – an exaggerated ellipse, while the orbits of Venus and Neptune appear virtually circular.)

ECLIPSE   the disappearance of a part or all of one heavenly body behind another, or in the other's shadow. (In a solar eclipse, the Moon blocks the Sun from view; in a lunar eclipse, Earth's shadow falls on the Moon.)

ECLIPTIC   the apparent path of the Sun, Moon and planets as seen from Earth, so named for the eclipses that occur here; the plane of the Zodiac and of Earth's orbit.

ELECTROMAGNETIC RADIATION   light, in all its guises, ranging from high-energy gamma rays and X-rays through ultraviolet radiation, visible light and infrared, to microwaves and radio waves.

ELONGATION   the most favourable time to view Mercury or Venus, the planets interior to Earth, when they achieve their greatest apparent distance east or west of the Sun. The greatest possible elongation for Mercury is 28 degrees, and for Venus, 47 degrees.

EPHEMERIS   a published table of predictions of heavenly bodies' positions, especially those of planets and comets.

ESCAPE VELOCITY   the speed a rocket (or any object) must

attain to break free of the pull of gravity at a planet's surface, and rise into space.

EXTREMOPHILE   any inhabitant of an extreme environment that would be toxic or otherwise unfit for all but properly adapted life forms.

GALAXY   a collection of billions of stars, all gravitationally bound, as in the Solar System's home galaxy, the Milky Way.

IGNEOUS   a term used to describe rocks formed from once-molten magma or lava.

KUIPER BELT   a doughnut-shaped region beyond the orbit of Neptune, named for Gerard Kuiper, containing hundreds of thousands of icy planetoids. Some of these objects, when deflected by gravity or collisions into orbits that carry them close to the Sun, become the comets that return on regularly repeating schedules.

MAGNETIC FIELD   the region around a magnet, throughout which the magnet affects charged particles or other magnets. Many planets, such as Jupiter and Earth, behave as giant magnets and generate their own magnetic fields.

MAGNETOSPHERE   the invisible bubble of a planet's magnetic field, defining the limits of the field's sphere of influence.

MAGNITUDE   the brightness of a heavenly body, expressed as a number; *apparent* magnitude (the body's relative brightness as seen from Earth) may differ significantly from its *absolute* magnitude, or intrinsic brightness.

MANTLE   the middle substance of a planet, filling the space between the surface crust and the core of a terrestrial world, or the upper atmosphere and solid centre of a gaseous one.

METHANE   also known as marsh gas, the simplest compound of hydrogen and carbon.

METEOR   a 'falling' or 'shooting' star, i.e., the light from a space rock or bit of comet dust descending through Earth's atmosphere and becoming incandescent from the heat of friction.

METEORITE   a landed piece of a meteoroid.

METEOROID   a space rock or chunk of a planet adrift in space.

MOON   Earth's natural satellite and, by extension, a body in orbit around any planet or asteroid.

NEBULA   a blurry-looking heavenly object, such as the disk in which a star is born.

OORT CLOUD   a spherical region of the outer Solar System, beyond the Kuiper Belt, named for Dutch astronomer Jan Oort (1900–1992). Comets from the Oort Cloud follow extremely long-period orbits, and may leave the Solar System after one swing around the Sun.

PERIGEE   that part of the Moon's (or an artificial satellite's) orbit that brings it closest to Earth, at which point it travels fastest.

PERIHELION   a planet's or comet's (or Sun-orbiting spacecraft's) closest approach to the Sun, and therefore the time of its greatest orbital velocity.

PLANET   a heavenly body, generally but not necessarily

expected to be larger than one thousand miles in diameter, and orbiting a star.

PLANETESIMAL   a chunk of material smaller than a planet, and which may join with other like pieces to become a planet or moon.

REGOLITH   dusty and rocky debris coating the surface of a terrestrial planet or satellite, similar to soil but lacking any live components.

ROCHE ZONE   the region close to a planet where tidal forces prohibit the build-up of planetesimals into satellites, named for French mathematician Edouard Roche (1820–1883), who first described it.

SATELLITE   a natural satellite is a moon; an artificial satellite is a spacecraft in orbit around a planet.

SOLSTICE   either of the two days each year (in June and December) when the Sun reaches its furthest distance above or below the equator, resulting in the shortest or the longest day.

STAR   a ball of gas, mostly hydrogen and helium, massive enough to ignite thermonuclear fusion at its core, and shine by its own emitted light.

SYZYGY   the all-in-a-line arrangement of heavenly bodies, such as the Sun, Moon and Earth during an eclipse, or the Sun, Venus and Earth during a Transit of Venus.

TESSERA (plural TESSERAE)   extremely deformed and fault-scarred areas that constitute the second most common land form on Venus (after volcanic plains), from the Latin word for 'tile'.

243

TRANSIT   the passage of one heavenly body in front of another, as when Mercury or Venus is seen passing across the disk of the Sun. The satellites of Jupiter and Saturn can also be observed in transit across their parent planets.

ZODIAC   the circle of twelve constellations through which the Sun seems to pass as the Earth makes its annual journey. These constellations correspond to the astrological signs of the zodiac: Aries, Taurus, Gemini, Cancer, Leo, Virgo, Libra, Scorpio, Sagittarius, Capricorn, Aquarius, Pisces.

DETAILS

1 MODEL WORLDS (*Overview*)

Model Solar Systems big enough to walk or drive through can be visited in Aroostook County, Maine; Boston, Massachusetts; Boulder, Colorado; Flagstaff, Arizona; Ithaca, New York; Peoria, Illinois; Washington, DC; Stockholm, Sweden; York, England; and in the Alps near St.-Luc, Switzerland.

The Soviet spacecraft *Venera 4* made the first probe of the Venusian atmosphere in 1967; *Venera 7* landed on Venus in 1970, and *Venera 8* in 1972. In November 1971, America's *Mariner 9* became the first Mars orbiter – the first spacecraft to orbit a planet beyond the Earth–Moon system. The Soviet *Mars 3* lander arrived the following month, but survived only twenty seconds on the Martian surface.

Michel Mayor and Didier Queloz of the Geneva Observatory made the first discovery of an exoplanet, and announced their findings about 51 Pegasi in October 1995. Two Americans – Geoffrey W. Marcy, University of California at Berkeley, and R. Paul Butler, now at the Carnegie Institution in Washington, DC – quickly confirmed the Swiss claims and went on to identify other extrasolar planets.

2 GENESIS (*The Sun*)

The extraordinary phenomenon of hydrogen fusion requires the tremendous heat and pressure found inside stars. Under normal circumstances on Earth, two hydrogen nuclei would never unite with one another, because both carry positive charge, and the electromagnetic force that causes two positively charged particles to repel each other is stronger than gravity. Inside the Sun, in contrast, high temperature pushes particles together so hard and fast that they collide despite electromagnetic repulsion. And once the particles are that close together, they succumb to a third force – called the 'strong force' because it is the strongest known in nature – which binds them together. The great power of the strong force, however, operates only over the tiniest distances, such as the size of an atomic nucleus.

In a single second inside its core, the Sun converts 700 million tons of hydrogen to 695 million tons of helium. The five-million-ton difference between input and outcome is transformed into the energy of light. This is a great deal of energy, according to the formula that describes energy (E) as the equivalent (=) of a given mass (m), or 5 million tons in this case, multiplied by the speed of light (c) squared (^{2}). Since the speed of light is a very high number to begin with (186,000 miles per second), squaring it – multiplying it by itself – yields a truly astronomical figure (34,596,000,000) that indicates the phenomenal power lurking inside even the tiniest amounts of matter.

Helium, the second most common ingredient (after hydrogen) in the Sun and throughout the universe, accounts for 10 per cent of the Sun's makeup. All other elements detectable by analysis of the Sun's light – carbon, nitrogen, oxygen, neon, magnesium, silicon, sulphur and iron, taken together – total only 2 per cent of the Sun's mass.

During periods of high solar activity, conglomerations of dark sunspots on the Sun dim its radiation by a few measurable tenths of a per cent, but overall the Sun stays a constant source of steady light.

The Moon at apogee (its greatest distance from Earth) cannot completely cover the Sun, but instead produces an 'annular' eclipse, in which the Sun appears as a glowing ring around the Moon and the corona may not be visible.

Although it is safe to look at the Sun during totality, the stages of partial eclipse preceding and following totality require eye protection.

3 MYTHOLOGY (*Mercury*)

Procrustes gained notoriety by lopping off his tall guests' legs and stretching short visitors on a rack to make them fit his bed, thus lending his name to violently or arbitrarily enforced conformity.

Mercury, travelling an elliptical orbit, reaches its peak velocity of thirty-five miles per second at perihelion, when it approaches within twenty-nine million miles of the Sun, and slows to twenty-four miles per second at the opposite orbital extreme, or aphelion, where the Mercury–Sun distance exceeds forty-three million miles.

The first of several mentions of 'Rosy-fingered dawn', as Homer called the reddish morning sky, occurs in Book I of the *Iliad*, line 477.

Transits of Mercury occur approximately thirteen times per century. Although the planet passes between the Earth and the Sun about four times a year, it most often travels above or below the Sun, when no transit is seen.

Mercury's period of spin is exactly two-thirds its orbital period, 'coupling' the two time intervals in a ratio of 3:2, or three rotations for every two orbits. (The discovery of the actual rotation rate was made by bouncing radar from the Arecibo Observatory in Puerto Rico off the surface of Mercury.) Most other tidally bound bodies in the Solar System display a 2:1 spin–orbit resonance. The most notable exception is the Moon, which completes one rotation per revolution about the Earth, giving it a 1:1 resonance.

4 BEAUTY (*Venus*)

William Blake wrote his ode to Venus in 1789, long before the discovery of the planet's own westerly winds. His mention of 'thy west wind' refers to evening breezes timed to her appearance.

Former President Jimmy Carter, while serving as governor of Georgia, reported Venus to the state police. During the Second World War, a squadron of B-29 pilots mistook the planet for a Japanese plane and tried to shoot it from the sky.

Donald W. Olson and Russell Doescher of Southwest Texas State University in San Marcos took their honours astronomy class to France in May 2000, and successfully identified the building featured in 'White House at Night' by using planetarium programmes to recreate the sky over France in the summer of 1890, reading letters Van Gogh wrote during his last weeks, and consulting archived weather reports.

The duration of a solar day on Venus, measured from one noon to the next, is 117 Earth-days, so that periods of light and dark last nearly fifty-nine Earth-days each. The sidereal day, or the time it takes the planet to rotate with respect to the background stars, is 243 Earth-days – longer than the Venus orbital year of 225 Earth-days. On Earth, as on Venus, the length of the solar day differs from the sidereal day; in Earth's case the solar day is about four minutes longer than the sidereal.

A complete Venus cycle – from morning star apparition to disappearance behind the Sun, through evening star apparition and disappearance in front of the Sun – lasts 584 days. This time period formed the foundation of the Mayan calendar. Since Venus makes eight orbits of the Sun in five Earth years, and passes between Earth and Sun five times in the process, there are five distinct 584-day Venus patterns in Earth's sky. The Mayas had a name for each.

Since 1919, authority for planetary nomenclature has been

vested in the International Astronomical Union. Although discoverers may suggest names for new satellites or other bodies, the choices must be approved by task and working groups, and ultimately voted into effect by the IAU General Assembly, which meets every three years.

5 GEOGRAPHY (*Earth*)

Even before Ptolemy, mapmakers applied concepts of latitude and longitude to the heavenly sphere and the globe of the Earth. After Ptolemy introduced a uniform coordinate system expressed in degrees, the ability to *determine* longitude awaited the late seventeenth century, and remained a problem at sea for another hundred years.

Ptolemy's *Geography* survived in manuscripts copied by scribes. The oldest such extant manuscript dates to the thirteenth century.

In 1828, in his *History of the Life and Voyages of Christopher Columbus*, American author Washington Irving popularized the romantic image of Columbus fighting for the roundness of the world. Medieval knowledge of the world's shape is well documented, however, in texts such as the thirteenth-century *Sphere* of Sacrobosco, and the world globe Martin Behaim completed months before Columbus left Spain. The Ancients could have concluded a round world from the stars visible at different latitudes, or the curved shape of the Earth's shadow on the Moon during a lunar eclipse.

Amerigo Vespucci's analysis of competing Portuguese and Spanish claims helped him estimate Earth's circumference at 27,000 Roman miles – just fifty modern miles shy of today's accepted value.

Earth's water supply constitutes only one-tenth of 1 per cent of the planet's mass, while outer Solar System moons such as

Ganymede, Callisto and Titan consist of 50 per cent water, most of it frozen.

After the next transit of Venus, predicted for 6 June 2012, there won't be another pair until 11 December 2117 and 8 December 2125. Transits occur in June or December because Earth crosses the plane of Venus's orbit in those months.

6 LUNACY (*The Moon*)

A 'blue Moon', widely reported to be the second full Moon in a calendar month, is more correctly (according to the 1937 *Maine Farmers' Almanac*, which defined the term) the third full Moon in a season that contains four of them. The *Almanac* reckons seasons by the tropical year, which begins on the day of the winter solstice, or 'Yule' (22 December). A true blue Moon, therefore, can occur only in the months of February, May, August and November.

Under a full Moon, a black-and-white landscape may reveal the greenness of grass, because the human retina is particularly sensitive to yellow-green wavelengths (the light the Sun emits most strongly).

Giovanni Riccioli (1598–1671), a Jesuit priest, established the system of lunar nomenclature still in use today. He and other selenographers (Moon mappers) named the mountains for Earthly ranges such as the Alps, Apennines, Caucasus and Carpathians. Craters on the Moon's near side honour great natural philosophers, from Plato and Aristotle to Tycho, Copernicus, Kepler and Galileo. Russian names apply to the far side, which was first imaged in October 1959 by the unmanned Soviet spacecraft *Luna 3*.

The Moon's rotation rate equals its revolution – 27.3 days – but by the time the Moon travels around the Earth to reach the point it started from, vis-à-vis the stars, the Earth has also moved. Thus the Moon is seen to require 29.5 days to complete an Earthly revolution and go through all its phases from one full Moon to the next.

7 SCI-FI (*Mars*)

Meteoriticist Roberta Score of the US Antarctic Program, in Denver, found the Mars rock known as ALH84001 on 27 December 1984. Scientists have successfully hunted meteorites in Antarctica since 1969. Analysis of ALH84001 began in mid-summer 1988 and tests confirming its Martian origin were completed by autumn 1993.

The hills near the Mawson and Mackay Glaciers, where the Mars rock was found, were mapped in 1957–8 and named for Professor R. S. Allan of the University of Canterbury, New Zealand.

The so-called 'Face on Mars', a topographical feature widely perceived to resemble a human face, appeared in *Viking* orbiter photos from 1976. Several media promulgated the suggestion that the face was an alien artefact, until subsequent imaging by the *Mars Global Surveyor* destroyed the illusion.

Giovanni Schiaparelli found what he called *canali* on Mars in 1877, eight years after the completion of the Suez Canal. Schiaparelli, trained as a hydraulic engineer, thought the straight lines no more the product of artificial intelligence than the English Channel, but later changed his mind. When Schiaparelli's sight failed, Percival Lowell took over observations – and interpretations – of the canals.

Johannes Kepler first imagined two moons for Mars in 1610, but the moons were not seen until August 1877, when Asaph Hall, working at the US Naval Observatory in Washington, DC, found them orbiting so close to the planet as to be nearly lost in its glare. He named them after two characters from Greek mythology, Phobos and Deimos, who were variously described by Homer as the sons of the war god Ares, or his attendants – or the horses that pulled his chariot.

8 ASTROLOGY (*Jupiter*)

Two surviving natal horoscopes drawn for (and probably *by*) Galileo are reproduced in vol. XIX of his complete works. An adept astrologer, he would not have classified individuals by Sun sign, as that practice arose in the twentieth century. Defining elements in the astrology of his time included the *horoscopus* (rising sign), the mid-heaven, the *immum coeli* (opposite of the mid-heaven), and the descendant sign on the chart's western horizon.

My interpretation of Galileo's natal chart is based on a reading by astrologer Elaine Peterson, 14 August 2003, and supplemented by listings in *The Complete Astrological Handbook* (see Bibliography).

Galileo's quote about 'fate' is taken from his *Starry Messenger*, in which he described his telescopic discoveries. Remarks directed to Cosimo come from the dedicatory introduction to that same book. Galileo's reference to the moons as 'stars' is appropriate terminology for his time, when 'the star of Jupiter' was seen as a rare 'wandering star' among the more numerous 'fixed stars' of the wider heavens.

After Galileo identified four Jovian moons in January 1610, no more were discovered until 1892, when Edward Barnard of the Lick Observatory in California found Amalthea. Another twelve surfaced in the twentieth century, four of them detected by *Voyager 2*. Names for these and another forty-three satellites detected recently by astronomers at the University of Hawaii continue the theme of Jupiter's intimates.

Henry Cavendish discovered hydrogen in 1766. Its metallic form, first predicted in the 1930s, was created at the Lawrence Livermore National Laboratory in California in 1996, by subjecting a thin film of liquid hydrogen to two million atmospheres of pressure.

The Sumerians of Mesopotamia recorded stellar observations as long ago as the eighteenth century BC. Several of their constellation names, including Leo and Taurus, are still used. The fully realized western zodiac dates from the middle of the fifth century BC.

Although the Jovian satellite Europa holds out hope of another abode of life within the Solar System, scientists feel certain the planet Jupiter is devoid of life. The *Galileo* probe found no complex organic molecules in its atmosphere.

9 MUSIC OF THE SPHERES (*Saturn*)

The Saturn of Greek mythology, called Cronus, devoured his children for fear they would kill him, as he had killed his own father, Uranus, to wrest control of the heavens. The infant Zeus (Jupiter), who escaped being devoured, later overthrew Cronus.

Saturn's so-called 'classical' rings – A, B and C – extend to a distince of 85,000 miles from the centre of the planet, or 170,000 miles across from tip to tip. These are the rings seen through a small telescope, and pictured in familiar images of Saturn. The narrow and twisted F ring, immediately exterior to the A, lies 2,000 miles beyond the perimeter of the A ring and its core is only 30 miles wide. The outlying, diaphanous E ring, which begins a little more than 100,000 miles from the planet's centre, is itself nearly 200,000 miles in width, so that its ring span of 600,000 miles more than doubles the distance from Earth to the Moon. It encompasses the orbit of the moon in Enceladus, and consists of icy debris that the shiny satellite sheds in its wake.

The D and E rings were detected by ground-based telescopes in 1966 and 1970, respectively. (E was actually discovered first, but astronomers questioned its reality for years, while D met a ready welcome.) *Pioneer 11* found the contorted F ring in 1979 and *Voyager 1* the G ring in 1980.

The Roche limit applies to objects held together by gravity. The *Cassini* spacecraft can dip safely inside Saturn's Roche zone because its parts are held together by nuts, bolts and the crystal cohesion of its metal molecules.

Resonant orbits, such as the 2:1 relationship between the Cassini Division and the moon Mimas, were first proposed in 1866 by Daniel Kirkwood, an American astronomer who used the resonance concept to explain gaps in the distribution of orbits in the Asteroid Belt.

Rotation periods of the giant planets were originally gauged by timing the reappearance of distinctive storms. Now they are determined by the rotation rate of each planet's magnetosphere, as measured by *Voyager 2*. Since a planet's magnetic field arises deep in the interior, scientists assume the two spin together at the same rate.

10 DISCOVERY (*Uranus and Neptune*)

The epigraph in italics is taken from one of Maria Mitchell's lectures, published posthumously by her sister Phebe Mitchell Kendall.

For this chapter, I assumed Maria Mitchell wrote of her 1847 find to the only other woman in the world who had discovered a comet, Caroline Herschel (1750–1848). In composing Miss Herschel's reply, I 'fictionalized' only the form, not the factual material. Miss Herschel was her brother's assistant when he discovered Uranus. At the time of Neptune's discovery, she was still active and intellectually engaged, despite her ninety-six years, and received word of the new planet from explorer Alexander (Baron von) Humboldt. Miss Herschel's correspondence put her in touch with most leading figures in this phenomenal epoch in the history of astronomy, and she met many of them in person, including King George III, his royal family, and three of his Astronomers Royal, as well as Giuseppe Piazzi (discoverer of the first asteroid), Carl Friedrich Gauss and Johann Encke.

The autumn 1847 discovery of Comet Mitchell preceded Miss Herschel's death by three months. Miss Mitchell worked then as librarian of Nantucket Island, and lived with her family in an apartment over the bank, of which her father was president. William Mitchell, a serious amateur astronomer, had built an observatory on the bank's roof, where he and she spent much time. In recognition of her discovery, Miss Mitchell won a gold medal from the King of Denmark, a $100 prize from the Smithsonian Institution, and election to honorary membership in the American Academy of Arts and Sciences. Later she became the first professor of astronomy at Vassar College, and led student expeditions to two total solar eclipses. On her 1857–8 trip to Europe, when she stayed at the home of Sir John and Lady Herschel, they gave her a page from one of the notebooks 'Aunt Caroline' had used to record Sir William's observations.

The biographical footnotes giving astronomers' life dates indeed support Miss Mitchell's prescription of 'night air' for longevity.

Whenever Sir William polished a telescope mirror, Caroline Herschel says in her *Memoir*, 'by way of keeping him alive I was constantly obliged to feed him by putting the victuals by bits into his mouth'. She did not mind such tasks: 'When I found that a hand was sometimes wanted when any particular measures were to be made with the lamp micrometer, &c., or a fire to be kept up, or a dish of coffee necessary during a long night's watching, I undertook with pleasure what others might have thought a hardship.' Her labours sometimes proved arduous: 'The mirror was to be cast in a mould of loam prepared from horse dung, of which an immense quantity was to be pounded in a mortar and sifted through a fine sieve. It was an endless piece of work, and served me for many an hour's exercise.'

The first five known moons of Uranus are Sir William's Oberon and Titania, the slightly dimmer Ariel and Umbriel, first seen by William Lassell from Liverpool in 1851, and Miranda, the nearest to Uranus, as well as the brightest and smallest, discovered in 1948 by Gerard Kuiper, and named by him for the heroine of *The Tempest*.

Sir John Herschel must have been thinking generally of sprites and sylphs in English literature when he named the first four Uranian moons, for Umbriel (like the later Belinda) belongs to 'The Rape of the Lock' by Alexander Pope. After Kuiper added Miranda, Shakespeare dominated subsequent choices. Five moons, spotted since 1997 with the Hale Telescope in California, honour Miranda's father, Prospero, and *Tempest* characters Caliban, Stephano, Sycorax and Setebos.

The planetary interiors of Uranus and Neptune evoke the 'hot ice and wondrous strange snow' in *A Midsummer Night's Dream* (V, i):

> A *tedious brief scene of young Pyramus,*
> *And his love Thisbe; very tragical mirth.*
> Merry and tragical! tedious and brief!
> That is, hot ice and wondrous strange snow.
> How shall we find the concord of this discord?

Following the discovery of Uranus's rings in 1977 by James Elliot of MIT and his colleagues aboard the Kuiper Airborne Observatory, *Voyager 1* saw evidence of faint rings at Jupiter in March 1979. Its sister ship, *Voyager 2*, confirmed the discovery three months later.

Neptune's ring arcs are named for Adams, Leverrier, Galle, Lassell and François Arago (the leading French astronomer who urged Leverrier to study Uranus), but there is none for Airy.

11 UFO (*Pluto*)

A heavenly body's motion against the background of the fixed stars reveals the object to be a wanderer of some kind, whether a planet, a comet, or an asteroid. The day-to-day shift in position, as noted in written records or caught on a sequence of photographic plates, is a parallax effect created by the Earth's motion. Tombaugh studied his photographic plates with a blink comparator – an instrument that automatically blinked back and forth between magnified views of the same region of space taken at different times.

The Lowell Observatory withheld announcement of Planet X's detection until 13 March 1930, to coincide with what would have been Percival Lowell's 75th birthday, as well as the 149th anniversary of the discovery of Uranus. Mrs Lowell, the former Constance Savage Keith, selected the name 'Zeus' for the new planet, then changed her mind to 'Percival', and finally to 'Constance', but the Observatory staff preferred the name suggested by eleven-year-old Venetia Burney of Oxford, England, and communicated to them by cable. 'Pluto' not only fitted the mythological scheme of planetary names (and had figured in the staff's top three picks even before the cable arrived), but also commemorated the founder's initials, 'P. L.'

Counting the Earth–Sun distance as one astronomical unit (AU), Jupiter is stationed at 5 AU and Neptune at 30, while Pluto and more than one hundred other members of the Kuiper Belt travel between 30 and 50 AU. The 17-degree tilt of Pluto's orbit carries it by turns 8 AU above the plane of the Solar System and 13 AU below. The actual distance between Pluto and Neptune remains at least 17 AU at all times because of the stable resonance of their orbits.

James W. Christy and Robert S. Harrington, of the US Naval Observatory in Washington, DC, deduced the presence of Charon from images of Pluto taken at Flagstaff, Arizona, only a short distance from Mars Hill. Christy named the moon for his wife, Char (short for Charlene), and also for the boatman Charon who ferried dead souls across the River Styx into Pluto's underworld.

David Jewitt (Institute of Astronomy, Hawaii) and Jane Luu

257

(University of Leiden), while working together at the University of Hawaii's telescope on Mauna Kea, discovered the first Kuiper Belt Object, which they called 'Smiley', after the spy in the novels of John le Carré, though its official name remains 1992 QB1. Quaoar, Varuna and Ixion have all been discovered from Mount Palomar in California by the team of Mike Brown (Caltech), Chad Trujillo (Gemini Observatory) and David Rabinowitz (Yale), who have chosen their KBO names, following IAU guidelines, from among the worldwide catalogue of underworld deities.

Gerard Kuiper based his prediction of what is now called the Kuiper Belt on the motions of short-period comets such as Comet Halley and Comet Encke. Calculated orbits for these bodies suggested they originated in the Kuiper Belt region, and return to it whenever they disappear from view. In 1950, the same year Kuiper published this idea, Dutch astronomer Jan Oort used a similar argument to predict another, more distant reservoir of comets at 50,000 AU. While the Kuiper Belt is shaped as a torus (doughnut), the 'Oort Cloud' forms a spherical shell. The orbits of short-period comets from the Kuiper Belt rarely incline more than twenty degrees from the plane of the ecliptic. Long-period comets from the Oort Cloud, on the other hand, may travel paths of any inclination, even perpendicular to the ecliptic.

In Lowell's day, the Observatory on Mars Hill owned a cow, named Venus. After the ninth planet was discovered, Walt Disney appropriated the name Pluto for the cartoon dog he introduced in 1936. Clyde Tombaugh understandably chose that same name for his cat.

BIBLIOGRAPHY

Titles listed here are sources of scientific, historical and literary background. Current information about the planets unfolds as news reports posted in scientific journals and on the Internet, including the web pages of NASA (www.nasa.gov), the Space Telescope Science Institute (www.stsci.edu), the Planetary Society (www.planetary.org), the United States Geological Survey (http://planetarynames.wr.usgs.gov/), and the European Space Agency (www.esa.int).

Abrams, M. H., with E. Talbot Donaldson, Hallett Smith, Robert M. Adams, Samuel Holt Monk, George H. Ford and David Daiches, eds. *The Norton Anthology of English Literature.* 2 volumes. New York: Norton, 1962.

Ackerman, Diane. *The Planets: A Cosmic Pastoral.* New York: William Morrow, 1976.

Albers, Henry, ed. *Maria Mitchell: A Life in Journals and Letters.* Clinton Corners, NY: College Avenue Press, 2001.

Andrewes, William J. H., ed. *The Quest for Longitude.* Cambridge, Mass.: Collection of Historical Scientific Instruments (Harvard University Press), 1996.

Asimov, Isaac. *Asimov's Biographical Encyclopedia of Science and Technology.* New York: Doubleday, 1972.

Aveni, Anthony. *Conversing with the Planets.* New York: Times Books, 1992.

Barnett, Lincoln. *The Universe and Dr Einstein.* 2nd revised edition. New York: William Morrow, 1957.

Beatty, J. Kelly, with Carolyn Collins Petersen and Andrew Chaikin, eds. *The New Solar System.* Fourth edition. Cambridge, Mass.: Sky Publishing, and Cambridge, England: Cambridge University Press, 1999.

Bedini, Silvio A., Wernher von Braun and Fred L. Whipple. *Moon: Man's Greatest Adventure.* New York: Abrams, 1970.

Bennett, Jeffrey, with Megan Donahue, Nicholas Schneider and Mark Voit. *The Cosmic Perspective.* Third edition. San Francisco: Pearson/Addison Wesley, 2004.

Benson, Michael. *Beyond: Visions of the Interplanetary Probes.* New York: Abrams, 2003.

Boyce, Joseph M. *The Smithsonian Book of Mars.* Washington and London: Smithsonian Institution, 2002.

Bradbury, Ray. *The Martian Chronicles.* New York: Doubleday, 1950.

Breuton, Diana. *Many Moons.* New York: Prentice Hall, 1991.

Brian, Denis. *Einstein: A Life.* New York: John Wiley & Sons, 1996.

Burroughs, Edgar Rice. *The Gods of Mars.* New York: Grosset & Dunlap, 1918.

Caidin, Martin and Jay Barbree, with Susan Wright. *Destination Mars.* New York: Penguin Studio, 1997.

Calasso, Roberto. *The Marriage of Cadmus and Harmony.* Translated from the Italian by Tim Parks. New York: Knopf, 1993.

Cashford, Jules. *The Moon: Myth and Image.* New York: Four Walls Eight Windows, 2003.

Caspar, Max. *Kepler*. Translated and edited by C. Doris Hellman. New York: Dover, 1993.

Chaikin, Andrew. *A Man on the Moon*. New York: Viking, 1994.

Chapman, Clark R. *Planets of Rock and Ice*. New York: Scribner's, 1982.

Cherrington, Ernest H., Jr. *Exploring the Moon through Binoculars*. New York: McGraw Hill, 1969.

Clark, Ronald W. *Einstein: The Life and Times*. New York: World, 1971.

Columbus, Christopher. *The Log of Christopher Columbus*. Translated from the Las Casas abstract by Robert H. Fuson. Camden, Maine: International Marine (McGraw Hill), 1987.

Cooper, Henry S. F. *The Evening Star: Venus Observed*. New York: Farrar, Straus and Giroux, 1993.

Darwin, Charles. *Voyage of the* Beagle. Edited by Janet Browne and Michael Neve. New York: Penguin, 1989.

Doel, Ronald E. *Solar System Astronomy in America: Communities, Patronage, and Interdisciplinary Science, 1920–1960*. Cambridge, England: Cambridge University Press, 1996.

Elliott, James and Richard Kerr. *Rings: Discoveries from Galileo to Voyager*. Cambridge, Mass.: MIT Press, 1984.

Finley, Robert. *The Accidental Indies*. Montreal: McGill-Queen's University Press, 2000.

Galilei, Galileo. *Sidereus Nuncius* or *The Sidereal Messenger*. Translated by Albert van Helden. Chicago: University of Chicago Press, 1989.

Gingerich, Owen. *The Eye of Heaven: Ptolemy, Copernicus, Kepler*. New York: American Institute of Physics, 1993.

———. *The Great Copernicus Chase and Other Adventures in Astronomical History*. Cambridge, Mass.: Sky Publishing, 1992.

Golub, Leon and Jay M. Pasachoff. *Nearest Star: The Surprising Science of our Sun*. Cambridge, Mass.: Harvard University Press, 2001.

Grinspoon, David Harry. *Venus Revealed*. Reading, Mass.: Addison-Wesley, 1996.

Grosser, Morton. *The Discovery of Neptune*. New York: Dover, 1979.

Hamilton, Edith. *Mythology*. Boston: Little, Brown, 1940.

Hanbury-Tenison, Robin. *The Oxford Book of Exploration*. Oxford, England: Oxford University Press, 1993.

Hanlon, Michael. *The Worlds of Galileo: The Inside Story of NASA's Mission to Jupiter*. New York: St Martin's, 2001.

Harland, David M. *Jupiter Odyssey: The Story of NASA's Galileo Mission*, Chichester, UK: Springer/Praxis, 2000.

Hartmann, William K. *A Traveler's Guide to Mars*. New York: Workman, 2003.

Heath, Robin. *Sun, Moon and Earth*. New York: Walker, 1999.

Herbert, Frank. *Dune*. Radnor, Penn.: Chilton, 1965.

Herschel, M. C. *Memoir and Correspondence of Caroline Herschel*. New York: Appleton, 1876.

Holst, Gustav. *The Planets in Full Score*. Mineola, N.Y.: Dover, 1996. (Score originally published by Goodwin & Tabb Ltd, London, 1921.)

Holst, Imogen. *Gustav Holst: A Biography*. London: Oxford University Press, 1938 and 1969.

——. *The Music of Gustav Holst*. London: Oxford University Press, 1951.

Howell, Alice O. *Jungian Symbolism in Astrology*. Wheaton, Ill.: Theosophical Publishing House, 1987.

Isacoff, Stuart. *Temperament: How Music Became a Battle-ground for the Great Minds of Western Civilization*. New York: Random House, 2001.

Johnson, Donald S. *Phantom Islands of the Atlantic: The Legends of Seven Lands That Never Were*. New York: Walker, 1996.

Jones, Marc Edmund. *Astrology: How and Why It Works*. Baltimore: Pelican, 1971.

Keats, John. *Complete Poetry of John Keats*. New York: Modern Library, 1951.

Kline, Naomi Reed. *Maps of Medieval Thought*. Woodbridge, England: Boydell, 2001.

Kluger, Jeffrey. *Journey Beyond Selene*. New York: Simon & Schuster, 1999.

Krupp, E. C. *Beyond the Blue Horizon*. New York: Harper-Collins, 1991.

Lachièze-Rey, Marc and Jean-Pierre Luminet. *Celestial Treasury*. Translated by Joe Laredo. Cambridge: Cambridge University Press, 2001.

Lathem, Edward Connery, ed. *The Poetry of Robert Frost*. New York: Holt, 1979.

Levy, David H. *Clyde Tombaugh: Discoverer of Planet Pluto*. Tucson: University of Arizona Press, 1991.

——. *Comets: Creators and Destroyers*. New York: Simon & Schuster, 1998.

Lewis, C. S. *Poems*. New York: Harcourt Brace, 1964.

Light, Michael. *Full Moon*. New York: Knopf, 1999.

Lowell, Percival. *Mars*. London: Longmans, Green, 1896. (Elibron Classics Replica Edition.)

Mailer, Norman. *Of A Fire On The Moon*. Boston: Little, Brown, 1969.

Maor, Eli. *June 8, 2004: Venus in Transit*. Princeton: Princeton University Press, 2000.

Miller, Anistatia R. and Jared M. Brown. *The Complete Astrological Handbook for the Twenty-First Century*. New York: Schocken, 1999.

Miner, Ellis D. and Randii R. Wessen. *Neptune: The planet, rings and satellites*. Chichester, UK: Springer-Praxis, 2001.

Morton, Oliver. *Mapping Mars*. London: Fourth Estate, 2002.

Obregón, Mauricio. *Beyond the Edge of the Sea*. New York: Random House, 2001.

Ottewell, Guy. *The Thousand-Yard Model* or *The Earth as a Peppercorn*. Greenville, SC: Astronomical Workshop, 1989.

Panek, Richard. *Seeing and Believing: How the Telescope Opened Our Eyes and Minds to the Heavens*. New York: Viking, 1998.

Peebles, Curtis. *Asteroids: A History*. Washington, DC: Smithsonian Institution, 2000.

Price, A. Grenfell, ed. *The Explorations of Captain James Cook in the Pacific as Told by Selections of his own Journals 1768–1779*. New York: Dover, 1971.

Proctor, Mary. *Romance of the Planets.* New York: Harper, 1929.

Ptolemy, Claudius. *Almagest.* Translated by G. J. Toomer. Princeton: Princeton University Press, 1998.

——. *Geography.* Translated by J. Lennart Berggren and Alexander Jones. Princeton: Princeton University Press, 2000.

Putnam, William Lowell. *The Explorers of Mars Hill.* West Kennebunk, Me.: Phoenix, 1994.

Rudhyar, Dane. *The Astrology of Personality.* Santa Fe: Aurora, 1991.

Sagan, Carl. *The Cosmic Connection: An Extra-terrestrial Perspective.* New York: Anchor Press, 1973.

——. *Pale Blue Dot: A Vision of the Human Future in Space.* New York: Random House, 1994.

Schaaf, Fred. *The Starry Room: Naked Eye Astronomy in the Intimate Universe.* New York: John Wiley & Sons, 1988.

Schwab, Gustav. *Gods and Heroes of Ancient Greece.* New York: Pantheon, 1946.

Sheehan, William. *Planets and Perception.* Tucson: University of Arizona Press, 1988.

——. *Worlds in the Sky: Planetary Discovery from Earliest Times through Voyager and Magellan.* Tucson: University of Arizona Press, 1992.

—— and Thomas A. Dobbins. *Epic Moon.* Richmond, Va.: Willmann-Bell, 2001.

Standage, Tom. *The Neptune File.* New York: Walker, 2000.

Stern, S. Alan. *Our Worlds.* Cambridge, England: Cambridge University Press, 1999.

——. *Worlds Beyond.* Cambridge, England: Cambridge University Press, 2002.

— and Jacqueline Mitton. *Pluto and Charon: Ice Worlds on the Ragged Edge of the Solar System.* New York: John Wiley & Sons, 1999.

Strauss, David. *Percival Lowell: The Culture and Science of a Boston Brahmin.* Cambridge, Mass.: Harvard University Press, 2001.

Strom, Robert G. *Mercury: The Elusive Planet.* Washington and London: Smithsonian Institution, 1987.

Thrower, Norman J. W., ed. *The Three Voyages of Edmond Halley in the* Paramore *1698–1701.* London: Hakluyt Society, 1981.

Tombaugh, Clyde W. and Patrick Moore. *Out of the Darkness: The Planet Pluto.* Harrisburg, Pa.: Stackpole, 1980.

Tyson, Neil de Grasse, with Charles Liu and Robert Irion, eds. *One Universe.* Washington, DC: Joseph Henry Press, 2000.

Van Helden, Albert. *Measuring the Universe.* Chicago: University of Chicago Press, 1985.

Walker, Christopher, ed. *Astronomy Before the Telescope.* London: British Museum, 1996.

Weissman, Paul R., with Lucy-Ann McFadden and Torrence V. Johnson, eds. *Encyclopedia of the Solar System.* San Diego: Academic Press, 1999.

Wells, H. G. *The War of the Worlds.* London: William Heinemann, 1898.

Whitaker, Ewen A. *Mapping and Naming the Moon.* Cambridge: Cambridge University Press, 1999.

Whitfield, Peter. *Astrology: A History.* New York: Abrams, 2001.

Wilford, John Noble. *Mars Beckons.* New York: Knopf, 1990.

Williams, J. E. D. *From Sails to Satellites: The Origin and Development of Navigational Science.* Oxford, England: Oxford University Press, 1992.

Wolter, John A. and Ronald E. Grim, eds. *Images of the World: The Atlas Through History.* Washington, DC: Library of Congress, 1997.

Wood, Charles A. *The Modern Moon: A Personal View.* Cambridge, Mass.: Sky Publishing, 2003.

Zubrin, Robert, with Richard Wagner. *The Case for Mars.* New York: Free Press, 1996.

ILLUSTRATIONS

196   Sir William Herschel's largest telescope.
Photograph © Bettmann/Corbis.

ENDPAPERS:
Details of 30 Doradus Nebula taken by Hubble.
Photograph © Corbis Sygma.